Science Versus Politics and Economics

WHY SCIENTIFIC LITERACY IS ESSENTIAL

by Leon R. McNarry

Produced by:

FriesenPress

Suite 300 – 852 Fort Street
Victoria, BC, Canada V8W 1H8

www.friesenpress.com

Distributed to the trade by The Ingram Book Company

Other books by Leon R. McNarry

The Sacred Fire

Poems Tales and Whimsy

For all of those whose young shoulders will have to bear the burden of society's intransigent mistakes.

Table Of Contents

Foreword Vii

Introduction Xi

Chapter One: The Mysticism Of Wordsworth—1946 1

Chapter Two: Science, Education And Catastrophe—1946 7

Chapter Three: Education For Tomorrow's World—1957 11

Chapter Four: CBC National Radio—Personally Speaking
—May 21, 1964 19

Chapter Five: The Challenge Of Change—1967:
A Panel On Industry Looks At Science In Our Schools 25

Chapter Six: That Which Is & My Convictions—1967 31

Chapter Seven: Science For Non-Science
Oriented Students—1968 35

Chapter Eight: Science, Students And Semantics—1971 41

Chapter Nine: Of Physics And Men—1972 67

Chapter Ten: Learning Processes And The Computer—1973 71

Chapter Eleven: Some Implications Of
Contracting-Out Research—1978 83

Chapter Twelve: Science, Technology And Society—1986 95

Chapter Thirteen: On Watching "Earth" Tvo,
November 24, 1987 111

Chapter Fourteen: The Existence Of God—1988 115

Chapter Fifteen: Does God Exist?—1988 129

Chapter Sixteen: Thinking, Emotions And Logic—1989 133

Chapter Seventeen: Concerns About Science,
Technology And Modern Society—1990 137

Chapter Eighteen: The Canadian Dilemma—Meech Lake—
1990 143

Chapter Nineteen: On The CBC Budget Cuts—1990 155

Chapter Twenty: Some Thoughts Following
The Constitutional Conferences—1992 163

Chapter Twenty-One: World-Wide Problems, Deficits
And The Universality Of Excessive Income—1992 171

Chapter Twenty-Two: TV Violence
And The Safety Of Children—1993 179

Chapter Twenty-Three: Politicians & The Science
Of Climate Change—2011 185

Postscript 193

Acknowledgment 195

About The Author 197

FOREWORD

THE MEANING OF THE BOOK of Knowledge—1911

Arthur Mee, Editor

I was seven years old when I was introduced to *The Book of Knowledge* in Grades Three and Four and again in Grades Five and Six at the Consolidated School District #525, in Foxwarren, Manitoba; there was a complete set of twenty volumes in those two classrooms. By the time I had finished Grade Six I had read most of those twenty volumes; that is where I received my real education about the world around me. Early in my life I realized the knowledge of *how* and *why* things worked was far more satisfying than not knowing; in other words ignorance is not bliss! While I was stationed at Rodel Park on the Isle of Lewis, off the coast of Scotland, during WWII, I heard of the death of Arthur Mee; it was a sad day as I mourned the loss of my unknown friend. *The Book of Knowledge* was first published as *The Children's Encyclopedia* in 1874; I eventually purchased our twenty volume, 1911 edition, in 1960 for $20. *Much of what Arthur Mee told us in 1911 is still applicable today.*

Arthur Mee explains the concept and purpose of the 1911 edition of The Book of Knowledge—his insights can be applied by today's teachers with profit for their students.

THE MEANING OF OUR BOOK

"This work is an attempt to bring together that part of human knowledge really worth while, so that even a child may understand. Nothing could be more false to its purpose than to imagine that it seeks to cram the mind with things that children need not know. It is based upon a definite idea of education which the editors have developed both from study and experience and which they have worked out in the volumes now before you. The bringing up of a child is conceived as the supreme task in which we can engage, but there is no sympathy with those who would set a child down at a desk almost before it can run. It is believed that in the early years a child is its own teacher; and that in a right environment it will teach itself more than all the teachers in all the schools can teach it.

It cannot be urged against this book, therefore, that it has come to steal away the joy of childhood and put a bitter grinding in its place. It has come, indeed, to bring more joy to childhood, believing that true joy of life comes from sympathy and understanding. Left to wander in this field, the child will find whatever it wants. For the youngest of all, its nurse will find her lullaby. The child in the nursery will find its nursery rhymes, and all the best stories that have ever been told. The child who can be left out of doors to play will find here the beginning of its interest in natural things. All the games and pastimes, all the fireside enjoyment the children love, the mechanical interests of boys, the domestic interests of girls, and home-made toys for both of them—this is but one phase of the practical value of the book. For the boy and girl at school these pages teem with precious things; for fathers and mothers, teachers and governesses, they may well become invaluable. It is a book for grown-ups and children too—to be read by children or to children. It is an encyclopedia of everything that comes into childhood and by childhood is meant all that period of life when the sensitive mind is being formed by the influences about it.

The *Book of Knowledge* is what it pretends to be. It is written in the words the children know. The art of saying things simply has long been dying out, but the writers of this book will seek to revive it. They will be simple by being natural; they will make a children's book without childishness, a book that children may read because it is simple, and that men may read because it is plain. The great mysteries will be made as clear as words can make them; and while the child will find its sense of wonder grow, it will find, too, that its mind is widening all the time, understanding more and more.

Note: The language may be quaint by today's standards—the use of the word *it* to designate a child—but his intent is clear. Mee was a devout Christian, but he lets the child make up her/his own mind about those values, while indirectly extolling those of integrity, honesty, and compassion. I expect that my father was instrumental in the purchase of those books for our school!

INTRODUCTION

This listing of my essays and writings is presented in chronological order to show the evolution of my concerns about the interaction of science with society. I am quite sure that the dawning of those concerns began as I read from the twenty volume set of Arthur Mee's *The Book of Knowledge* while at the public school in the Consolidated School District of Foxwarren, Manitoba. This quote from the Canadian School Board Journal, October, 1925, is an indication of the quality my early education.

CANADIAN SCHOOL B O A R D JOURNAL

Consolidated School at Foxwarren, Manitoba.

By J. H. Plewes, Superintendent

There is little necessity to-day for advancing arguments for Consolidated Schools as they have proven their value in over one hundred districts in Manitoba alone. Perhaps the best of these schools is Foxwarren situated near the western border of the province. On this point we may take the opinion of Dr. Works, Director of Rural Education in Cornell University, who pronounced it the best that he had seen anywhere.

...Some districts have been commended, and rightly so, for not erecting expensive school buildings in these hard years but when building becomes a necessity not only should the needs of the present be considered but the probable needs of the next fifty years. In this case the next generation will praise the foresight of the public spirited men and women who planned so wisely for them. (My father was one of them.)

(From the *Canadian School Board Journal*, Volume IV, No.11. Page 18. Published under the authority of the Canadian School Board Trustee's Association at Port Perry, Ontario, October 1925.)

I served in the R.C.A.F. as a Radar Officer during WWII; following my graduation from the University of Western Ontario in 1950 at age 34, I was employed by the National Research Council in Ottawa as a Research Scientist. I retired in December, 1979, as a Senior Research Scientist. It was during those early years that I began to write the essays which make up the body of this book.

A common thread flows through these essays; it is that the fundamental laws of nature comprise the ultimate reality within which we enact our lives. We may think and organize our lives as we wish, but, and it is a very big but, what we do, whatever its manifestation, is inevitably subject to those elemental laws. It is important to understand that those laws are immutable throughout the universe. When we discover them, they are attributable to the science of physics or chemistry. (Biology is ultimately determined by the laws of physics and chemistry.)

Love (not romantic love), compassion and empathy, leading ultimately to morality, are intertwined threads which have had a very long evolutionary gestation. Those qualities are the result of our *being* as biological animals, to whom we have given the designation *homo-sapiens*. The meanings attributed to *sapiens* (Latin for wise) need not concern us here—I generally regard the meaning to indicate self-awareness including the use of language. These threads have enabled the formation of societies.

Once societies, be they hunter-gatherer or modern-industrial, arise the organization of that society becomes important; they may be tribal with a chief or leader who controls the tribe, or as in modern societies, political with a formal constitution and its code of laws. My concern is that such societies (e.g. countries) have become so involved with their culture that they give little or no consideration to those basic fundamental laws that govern an uncomprehending Nature. It is we, the biologically evolved component, that has comprehension and hence the ability to acquire culture, and, in that process, our awareness of the nature of the Universe, as it applies to us, is generally unrecognized or ignored.

For societies to grow and function, accumulated knowledge is formulated and passed on to succeeding generation through the process of education from which formal structures have grown such as schools, colleges and universities. Few of us really understand the process by which we become educated. A very large part is absorbed, rather than taught. Currently, the electronic media and the world-wide Internet are major contributors. Educators themselves are usually unaware of the import and meaning implied in the second paragraph above.

My concerns have arisen from these common threads, which infuse my writings and essays. It is my hope they will give you pause to think on them, and to add your own thoughts and insights. Our interaction with the natural universal laws governing our little bit of the universe has now reached a point where that interaction has begun to react against us—we are now approaching a tipping point from which there is no return. Global warming and the general ignorance of the consequences arising from indiscriminate use of technologies are real, and will eventually become fatal to life on planet Earth! (February 2012)

CHAPTER ONE
THE MYSTICISM OF
WORDSWORTH—1946

This was an essay written as an assignment during my first year at the University of Western Ontario. Looking back, I can see the beginnings of my concerns about both spirituality and my understanding of the natural world around us. (References have been omitted.)

In considering the mysticism of William Wordsworth it is necessary to have a clear conception of two things; first, the definition of mysticism, and secondly, the influence of nature upon Wordsworth, particularly during his formative years. These two factors are complementary but initially will be considered separately to give perspective to the discussion.

A precise definition of mysticism is somewhat elusive for there are as many variations of mysticism as there are mystics. However, it has been defined as "a faith in the possibility of the union of the soul of man with some higher and greater spirit or force such as Nature or some conception of deity". It is to be noticed that mysticism is defined relative to nature. From early childhood Wordsworth spent much of his time in the hills and valleys surrounding his home. It is apparent that, while he did not then analyze his feelings, nature in all her various moods and forms had a profound influence upon his mind.

"To me the meanest flower that blows can give
Thoughts that do often lie too deep for tears".

He did not realize at the time the full impact of this influence, though the effect and impressions were vivid and disturbing.

"…huge and mighty forms, that do not live
Like living men, moved slowly through the mind
By day, and were a trouble to my dreams".

Wordsworth grew up in close association with nature throughout his boyhood. It permeated his soul and later in his writings he turned to these experiences for his thoughts and material. As he expressed it, "Fair seed time had my soul". The emotional intensity of feeling that Wordsworth experienced was almost trance-like; it was of these moments that he records in Tintern Abbey:

"…we are asleep
In body, and become a living soul:
While with an eye made quiet by this power
Of harmony, and the deep power of joy,
We see into the life of things".

In nature, Wordsworth has perceived a spirit or power which is "interfused" with and "rolls through all things". The question arises—what is the force or spirit in nature that moved Wordsworth so deeply?

Nature is the sum total of forces which animate the created world or the aggregate of events and changing things which make it up. All that exists is a manifestation of nature and therefore to nature may be attributed all that is perceived by the human mind. This is what Wordsworth meant when he wrote:

"Wisdom and Spirit of the Universe!
Thou Soul, that art the Eternity of thought!
And giv'st to forms and images a breath
And everlasting motion!"

Only through nature then can truth be known. Wordsworth through his mystic insight came to perceive and believe nature to be the source of all truth and knowledge. It was natural that he should attribute to nature the quality of deity or God.

Man has always sought the ultimate explanation of natural phenomena. In the earliest times these explanations were only mental or verbal pictures, later they were recorded on cave walls as Gods of the Elements. It is significant that peoples from the earliest times have attributed their idea of the Supreme Being to some phase of nature. Later the spread of knowledge, through books and schools, refined this seeking for the truth.

"Nature has indeed made knowledge possible. It provided certain animals, particularly human beings, with the superior ability to respond at a distance from stimulating agents, rather than on contact, and also in accordance with past experience and in advance of or in preparation for, stimulation". This theory places truth in the category of eternal existence and knowledge of the truth an attribute of man. Fundamentally, knowledge of the truth or deity can only be acquired from nature. It follows then that Wordsworth, through his contact with nature and its profound impression upon him, should have inclined his thinking along mystical lines.

Mysticism involves a perception of ideas through spiritual contact with some intangible superior being or force and this type of thinking often leads to an interpretation of nature in terms of Pantheism.

Wordsworth has said of himself:

> " ...I was often unable to think of external things
> as having an external existence, and I communed
> with all that I saw as something not apart from,
> but inherent in, my own material nature. Many
> times while going to school have I grasped at a wall
> or tree to recall myself from this abyss of idealism
> to the reality."

This passage shows how real to Wordsworth was his mystical interpretation of nature, yet he could not wholly believe for he "grasped at a wall" to recall himself to the "reality".

There was yet another troubling aspect of his mysticism. How could he reconcile all the apparent "evil tongues, rash judgments ...sneers of selfish men", with the goodness of the omnipresent deity? This question is fundamental to the conception of Pantheism. Wordsworth did not attempt a concrete answer to this problem, but, "knowing that Nature never did betray the heart that loved her", he let his faith comfort him, secure in his own trusting belief in her goodness.

No discussion of the mysticism of Wordsworth would be complete without reference to the influence his poetry has, or can have, upon present day thinking and philosophy. Today Wordsworth can be read for the deep, quiet beauty that emanates from his better works. It can also be read as a concept, then new, of the natural world; true, other mystics and philosophers had expressed similar ideas, but none so forcefully as did Wordsworth.

The trend in the modern world towards concrete, scientific explanations for all observed phenomena tends to induce a cynical attitude towards mysticism, particularly of the type expressed by Wordsworth. It is interesting to note that the majority of Christian doctrines are essentially mystic in character. Many of his interpreters and critics try to analyze what he has written and felt, in an attempt to prove or disprove his ideas. Such criticisms are of necessity based upon the writer's impression of what is truth. It can be stated with some degree of assurance that truth is an unalterable, eternal entity; whether man has interpreted what he perceives and feels correctly or not, does not alter the fundamental facts of the Universe. Modern science is tending more and more to show the oneness of all things physical. It may be that Wordsworth in his moments of mystical insight when he saw "into the life of things", expressed a truth that staggers the imagination of man, and thus causes his unbelief.

Whatever the interpretation of Wordsworth's mysticism may be, he has written poetry that is still a powerful influence today, as the volume of literature about his works testifies. He believed in nature without question as the source of spiritual growth, strength and solace for the troubled mind. He has shown us a way to nature if we will but go, where it is possible to:

> "Hold infinity in the palm of your hand
> And eternity in an hour".

CHAPTER TWO
SCIENCE, EDUCATION AND CATASTROPHE—1946

Public Speaking was a mandatory topic in the First Year English course for returning WWII veterans at the University of Western Ontario; this was an assignment in which we had to convince our audience of the importance of a topic which concerned us—the atomic bomb had just been used to destroy two cities in Japan.

H. G. Wells once remarked that the fate of the world depends upon a race between education and catastrophe; but in view of the atomic bomb he has more recently announced that catastrophe is obviously winning, since science and technology are developing more rapidly than man's ability to control them.

Even without consideration for the source of this statement, its implications are profound. Science, that branch of human behaviour which fundamentally seeks only to discover the truth, and technology, the art, usually industrial, of applying the fruits of scientific research in a manner which makes them of practical use in the broader field of human society, have been so misapplied during the past six years that thinking men have grave doubts for the future of our present civilization. True, this misapplication has been forced by the war to a much greater extent than at any previous time, but, it cannot be denied that certain branches of science have been deliberately harnessed, to develop methods of destroying that which has been so laboriously built up by the common working man.

Now, what can be done to intercept and avert the impending catastrophe, as Mr. Wells calls it? From his statement it would seem that education might be a solution.

Again a question might be asked; what is education? Education, in its broadest sense, might be defined as partaking of that portion of the accumulated knowledge of the centuries that you wish to apply to your particular mode of life. Knowledge by itself is of little use. It is also necessary to have the power of reasoning, that is, the faculty of observing a set of actions and circumstances and inferring a result. The result of this reasoning, however, will be of no real benefit unless it is governed intelligently. Intelligence implies the ability to profit by experience. This faculty now approaches wisdom, the culminating point in the development of the mind. Hence, wisdom is the ultimate objective of education.

To merely acquire an education in the ordinary sense is to go around the base of the hill, and to see only that which is near and of immediate importance; to acquire wisdom is to endure until the summit is reached. Here, the scene takes on perspective, the true values can be determined, and those things which will profit a man in the future are more readily discerned.

If this hypothesis seems trite in its implications, consider the possibilities of an unfettered knowledge directing the atomic forces we know to exist. This is the catastrophe of which H. G. Wells was speaking. On the other hand, wisdom, which implies honour, would direct these energies to the benefit of mankind without regard for race or creed.

Education then, is the thing of prime importance in the world today, even more pressing than the United Nations Organization or the Commission to Control Atomic Energy. Erudition, or book learning, is not sufficient; experience in the ways of human society and a considerable amount of unselfish, unbiased thinking in world affairs is necessary. Success or failure of legislatures, international conferences, world governments almost entirely depend upon the perspicuity of all peoples and not only on the sagacity of their leaders or representatives.

The intellectual material is available, but it must be developed. Modern society, as a result of technological developments moves at such a pace that the vast majority of people have all their time occupied in eking out their precarious livelihood, with little time left in which to consider whither they are going.

In concluding, may I quote two points of view, each of which has its merits: first, Professor Hill, Foreign Secretary to the Royal Society, London, England, speaking of scientific ethics said: "We scientists throughout the world must take the initiative in these matters. We must not leave it to others who certainly will do nothing about it. If we do, we and our civilization may perish together."

And, more optimistically, Melvin Rader of the Department of Philosophy, University of Washington, writes in the *Scientific Monthly*, "Science has contributed in two essential respects to the establishment of a better type of civilization, first in its content and second in its method. In content, science is providing a new view of the world and man's place in it, a realistic vision which emphasizes the interrelatedness of things and which thereby re-enforces the co-operative ideal of human life. In method, science involves a broad community interpretation, a world-wide fellowship of truth seekers, and it pre-figures the co-operative organization of mankind."

CHAPTER THREE
EDUCATION FOR TOMORROW'S WORLD—1957

I was now employed as a research scientist by the National Research Council in Ottawa. Mavis and I had built our own house near the village of Cumberland, 12miles east of Ottawa. We were actively engaged with the local Home and School Association activities; I was nominated as the first President. This was the talk I gave at the first general meeting in 1957.

If you were asked to define one basic or fundamental aim of education what would it be? What is it that you hope education will do for your children to enable them to develop their full potentialities as human beings? The Winston Dictionary defines education as "... the training of the mental and moral powers either by a system of study and discipline, or by the experience of life. Education comprehends all that we assimilate from the beginning to the end of our lives in the development of the powers and faculties bestowed upon us at birth." Our children are born with certain potentialities but unless these potentialities are developed they will become little else but savages. *(I would not use that term today!)* Due to the fact that they are immersed in our civilization—or more correctly, culture— they become civilized. When a man dies a little bit of civilization dies with him; when a child is born he has no civilization. He must acquire it.

So education is the process whereby we transmit to our children the knowledge and wisdom of the present and past generations, i.e., by this process we make our culture available to them. We cannot force our children and young people to absorb knowledge and wisdom. It must be made available and they must be helped in such a way that there is deeply instilled into each one an abiding desire to learn, to think, to do, and let us not forget, to respect the culture which we are passing on to them. We are continually being educated throughout our lives and mankind is continually adding to his fund of knowledge. Thus in order to advance the intellect of our children to the point where we say, "*O.K. you are on your own now.*"; it is necessary to select, formulate, and prepare from our cultural heritage that which we consider as necessary education for our children, in order that they in their turn may profit by, add to, and transmit to their children the knowledge and wisdom of civilization.

But civilization does not only imply knowledge and wisdom, it brings with it unsolved problems of the past ages, many of which we will pass on unsolved to our children, together with problems of our own making. One of the fundamental aims of education must be to prepare our children so that they can meet and, let us hope, solve these problems.

We might say then that, apart from preparing our children to make their way and earn their livelihood in our world, education for participation in the world of tomorrow also means that they learn to enjoy the achievements of man's intellect in all its forms of literature, music and art—all man's creations in beauty of sound, form, and colour. To create new things from the work of the past and present, to explore new frontiers—yes, even create them. Also education for tomorrow's world means preparing our children to face, and if humanly possible, to solve the problems that we and our forefathers will be leaving for them. Don't forget that the child of today will be facing these problems by 1980, just look back and see how close the 1930s are to you.

Now, what are some of the problems our children will be facing? Some of them come to mind immediately such as the problems of the atomic age, but let us look at some others. What are we to do about the rate of growth of the world's population? How are we to communicate our ideas and ideals to other people and how are we in turn going to understand theirs? What about the growing discrepancy between the rates of development of technology and social consciousness? Are the products of our technology adequately distributed? And what about the problem of leisure time? You may say these problems don't affect us here in Canada. No one will be so foolish as to start an atomic war—we have millions of acres of unpopulated land—we can talk to one another—we understand what is going on in other countries—we have cars and T.V. sets and none of us ever have enough leisure time!

I would like to suggest that this is not so. In fact the belief that, apart from atomic war, everything is alright and we have no major worries is a fallacy and in itself a problem.

In order to understand a problem it must first be formulated. This can only be done properly when one has complete information at one's disposal. However, let us take a brief look at some of the situations our children will have to face and from this decide on some fundamental aims in the education of our children.

It has been said that the population bomb is as great a threat to human survival as the atomic bomb—only its fuse is longer. A few figures will put the problem in perspective. The present world population is about 2,700,000,000 people. This figure represents one twentieth of all the people who have lived since there is a record of man's existence! Just think about that for a moment before reading on. This figure is increasing at the rate of nearly 2 percent per year or 7,000 every hour. In 50 years the population figure will be doubled and yet it is estimated that tonight one half of the world's population will go to bed hungry. This rate of increase cannot continue for long. Take a look (in any good modern atlas) at the areas of subsistence agriculture and compare it with areas of greatest population density—almost all the facts are there. What has been our attack on this problem? Improve food production?

In the western world the limit is not set by lack of knowledge but rather by lack of application of knowledge and even more the limit is economic, i.e. does the farmer produce only in order to satisfy his own economic wants; how much of our supplies does our method of distribution allow us to ship to areas of want? Look at a calorie intake map in your atlas. India's annual population increase is 5 million and yet it is on the verge of starvation. The average daily calorie intake is 1600 calories compared with 3000 calories in the western hemisphere. What has the use of the "wonder drugs", malaria control, etc., done in underdeveloped countries? Ceylon's death rate has been decreased from 20.3/thousand to 12.6/thousand in 3 years. At this rate its population will double in about 25 years and even now it produces only about 1/2 of its food needs. Is the problem death control vs. birth control?

All the things about us are the products of our technology or industrial "know-how" as it is sometimes called. There seems to be no limit to the variety and numbers of things we can produce for our own consumption and for export to the rest of the world. There are two basic methods by which goods are distributed; direct barter and the use of a medium of exchange, i.e., money. Out of this has grown our capitalist private enterprise system. But does it work? Can it keep pace with modern technical expansion? Is it necessary to have boom and bust? Let us remember that it is we, the people, who live in this land and make use of its resources and that we make the decisions, through our elected representatives that govern the methods by which we process and distribute the wealth of the land and sea around us. The economic systems that we use to distribute our collective technological and agricultural production have evolved over a very long time and they are in a continuous state of flux when viewed in the long term. The direction can and must change, but how it is changed depends upon the intelligence our children bring to bear on the problem.

One fundamental problem in formulating and solving the world's problems is that of communication of ideas. One of the largest stumbling blocks is fear—fear of other countries whose language is different, whose ideas cannot be conveyed to the people as a whole— ideas communicated through translation, interpreters, etc., are undoubtedly coloured often deliberately to suit political expediency. Perhaps we should start by trying to understand our neighbours, their customs and cultures and to recognize them as fellow human beings who can feel joy and also be hungry, who can love their children and yet kill men in defending their way of life—who is to say that their ideologies and beliefs are inferior to ours—we are all members of the human race and the artificial barriers created can only serve to prolong our insecurity and problems.

Technology has provided us with means of instantaneous communication of ideas and very rapid means of transportation of goods and people; in other words the world has shrunk to an area which to our grandfathers would be represented by a few tens of miles.

And now what about leisure time? This may seem like a ludicrous anomaly to you after the previous topics. But let us suppose for a moment that the foregoing problems are solved. Man will have virtually unlimited energy at his disposal. Science will develop new materials and techniques which will free man of drudgery. This means inevitably that man will work less and produce more. Now the average man in North America works a 40-44 hour week, i.e., he works 5 days (farmers will laugh here!) and in the future he may well work only 3-4 days a week. This means he will have roughly one half of his time for leisure or pleasure. How will he use it? Creatively or destructively?

I believe that man is inherently a restive creature; otherwise he would still be living in caves from hand to mouth. Enforced leisure will then lead to mischief if man's energies cannot be used to satisfy this restive nature in a creative manner. Fortunately man has the ability to express himself in many ways. The general term I am thinking of here is art where art is defined in general as a skill developed as the result of knowledge and practice, and in particular when this skill is applied to painting, music, poetry, architecture, theatre, etc. Note that the products of these skills when recorded in their various forms have survived the ages. Hence they must have an intrinsic value to man. Please don't think that this applies only to so-called highbrow art—this is not the intention. Let us remember that these things and ideas are the products of the human mind and are an expression of its greatness. Are our children going to add to this heritage and in what manner? Are they going to appreciate and enjoy the creative efforts of their fellow man?

Conclusions and Summing Up

We have talked about the population bomb, our technology and our economic system, leisure time and art. Now what about education?

What I have done is to outline in extremely general terms what I consider to be major problems of an essentially global character that our children will inevitably face. Some of them threaten their very survival. Remember one definition of education; preparation for the work of life. Now from what has been said the work of life falls into two general categories, one involves our survival—the other our ability to enjoy the fruits of our labours and to partake of our culture. How are we to educate our children for their work of life?

It seems to me that the fundamental aims of education are twofold. Our children will have the immediate task of survival, i.e., they require knowledge and skills with which they can earn their livelihood. Secondly, they will have to chart their course in tomorrow's world and to do this they will have to think. This sounds like a dull platitude but what does the word think mean—to form in the mind—to conceive—to have in the mind as a notion, idea, etc. To "think" in the context I am using implies a reasoning process that is logical and concise; contrast with the term to "wonder" as it is often used. Also the context implies a moral outlook based upon broad human values which are common to all mankind—this has a broader base than the term "ethics", which can vary from culture to culture.

All man's actions and doings originate in his mind as thoughts. What he does depends upon how well he can marshal his facts and ideas, think on them and decide on a course of action. His thoughts are the result of past experience and influences as well as the present situation he is considering—whatever it may be. Today he is inundated with ideas from a multitude of sources, many of which are deliberately designed to influence the final outcome of his thinking—as I am doing now; for these reasons I believe that it is absolutely essential that our children learn the value of independent thought. In fact, I would go so far as to say that the survival of the human race depends upon this more than any other single thing. Only by thinking individually can we collectively meet and solve our problems. Only in this way can we stimulate our children to take an active rather than passive part in the progress of civilization.

So education for tomorrow's world, to me, means recognizing that our children are unique individuals and that their supreme creative faculty is their ability to think, and the fundamental aim of education is to develop this faculty.

CHAPTER FOUR
CBC NATIONAL RADIO—
PERSONALLY SPEAKING—MAY
21, 1964

This entry is derived from my broadcasts on the CBC national radio network. When I was interviewed prior to this broadcast, the local CBC Ottawa producer read my script and looked quite puzzled. Finally she looked at me and said, "Bertrand Russell can say this, but you can't!" (I took that as a left-handed compliment.) I was told to go back home and re-write the ten minute talk in my "own" words! This is the result. (I wish I could find that original script!)

The other day a physicist friend of mine phoned me; he had a lot to say about the teaching of science in our schools. He had been coaching a student who was having trouble with the concepts of mass and weight. The lad wrote an exam shortly afterwards and sure enough there was a question involving mass. He got zero on the question. Naturally, he tore straight over to my friend and demanded to know why he got zero when he had answered the question in the way my friend told him was correct. He had indeed done it correctly and was told to go back to the teacher and tell him that his answer was correct. After some discussion, the teacher gave him half marks. Now, I'll bet the lad is still confused, and my friend still has some strong things to say about the teaching of science in our schools.

Why did my friend phone me? Because several of my scientist friends and I have decided that it is high time that we—that is, the scientific community—came out of our ivory towers and gleaming labs to take a serious and responsible look at the teaching of science in our schools. Mention schools to a scientist parent, and you have an irate parent. But scientists have been irate about this for years, and have done little but sputter indignantly to their friends, and blame the science teachers.

Yes, I do have a daughter in high school. One evening, she was sitting at the kitchen table carefully and neatly drawing a diagram of a cell and labeling the parts: nucleus, cell wall, cytoplasm, vacuoles, etc. The cell was nice and regular; the nucleus was in the centre, and the vacuoles symmetrically placed in the corners. What caught my eye was not the unnatural symmetry but the term vacuoles. A recent issue of a science magazine was devoted entirely to the cell, but not a mention of vacuoles. With new and improved techniques much has been found out about the cell and vacuoles have gone by the board in the light of new knowledge. I gave this magazine to my daughter and she took it to school; two months later, I got it back—her science teacher had apparently been reluctant to return it until he had read it. Sure enough, on her last exam was the question, "Draw a diagram of a cell and label its parts". You guessed it—a neat, rectangular diagram with vacuoles—and full marks! The curriculum triumphs again! But again, is it fair to blame the teacher? *(I was wrong! Vacuoles are indeed part of the cell structure.)*

A few years ago a group of elementary school teachers in our rural school area decided to hold a science seminar. There are several scientists living in the area, so the teachers asked a couple of us to take part in the seminar, which was held at the local teachers' college. We were pleased to be asked, and did our best to help. The plea the teachers had was—where do we get authoritative information? How do we know we are teaching science that will satisfy scientists? We didn't seem able to help much at this science seminar. But the problem kept rankling in our minds. Did we have a responsibility here, or should we stick to our laboratories and let the educators cope with the problem? Would we merely be interfering in a situation that might be well in hand? But complaints from our scientist friends kept needling us. What do you say to someone who says "Know what they told my boy in grade seven last week? That evolution was just an idea cooked up by scientists who don't believe in God!" This father was fit to be tied!

What all this adds up to is that rising blood pressure never cures anything; but thoughtful, concerted action might. To get back to my friend who phoned me to ask, "Did I know what was going on to improve the teaching of science in our high schools?" I'm no expert in this field, but I do know that a great deal is being done, not only in Canada, but in the United States and elsewhere, particularly in the physical sciences, in biology, in chemistry and, of course, in mathematics; but somehow this effort is not reflected in our text books. Another local scientist who had spent a lot of time in Europe, said recently that the best science texts are Russian, German and French in that order—ours, he was sorry to say, were near the bottom of the list. If you remember, my friend was concerned about the confusion between mass and weight in the mind of the young lad he was coaching. So, I asked my daughter to bring home her grade nine science text book, and I had a look at the chapter dealing with measurements, and on page 49, I found this perennial statement, "Scientists call the measure of the amount of matter in an object its mass". This is not strictly true. Scientists ascribe to matter a property called mass, which makes itself evident by phenomena which involve gravity, inertia, and energy.

The text carefully points out that the student must not confuse weight and mass—good—but turn over the page, and we find a diagram of a triple beam balance, and a set of ten rules for the use of the balance. In every rule, the term mass, or masses, is used to denote particular objects or parts of the balance—never the term weight. But the diagram is labeled with terms such as 11 sliding weights". Weights, or masses, which is the student to accept?

In another chapter we are told, "In science, the mass of a given volume of a substance is called its density". Why, "In science"? Would this statement not be valid in an essay on science in an English text book? To make sure the student has got the idea, he is given problems to solve. In one problem the terms weight and mass are used interchangeably. It winds up by asking the student to find— quote "the mass of the bottle" unquote—when filled with a different liquid. Now, the mass of the bottle has not changed, no matter what common liquid it contains. The question is meaningless. But why go on? No wonder students have to ''unlearn'' what they are taught in high school. The concept of mass is not really all that difficult that a writer of a text book should display so much personal confusion. If this particular textbook had been edited individually by several competent scientists, such gross misunderstandings could have been cleared up. As it is, I know my daughter is confused, because, when I asked her what one means by mass, her rather dubious reply was, "It's how much something weighs, isn't it?" I doubt if many students could complete a course based on this text, without confusing mass, weight, volume and density.

It is my personal opinion that we will not have really first-rate textbooks and science curricula until the scientific community and the educators get together and try to understand each other's problems and viewpoints. With this in mind, two other physicists and I decided to hold a small meeting in Ottawa to which we invited interested scientists and educators to see if we could define at least the boundaries of the problem. What we were really doing was thinking aloud in public; we asked the local educators to sit in and hear what the scientists had to say, and then give them a chance to comment. The meeting turned out to be about twice as big as we had expected partly because of some excellent pre-coverage in the press, and partly because of a deep-seated concern about education on the part of local scientists.

When the meeting was over, we realized two things. Firstly, there was a genuine feeling of rapport between the scientists present, and the local educators; as one science teacher put it in a note of thanks, "It was a heart-warming experience to receive a few words of encouragement from the scientists at the meeting. This, along with a real understanding of the kind of problem a teacher has to meet, was given to me by you and your colleagues" and, secondly, we realized that we had a huge job ahead of us just to understand what was being done, not only here in Ottawa, but elsewhere. We do feel that we have established the principle that the scientific community; and by this I don't mean only the university professors, but the practic-ing scientist, who is the end-product of the whole range of science education, have a serious and responsible role to play in education.

I think most of us are aware that science and technology have a profound influence on our society and that the educators cannot carry the burden alone of evolving adequate science teaching. The scientists must help. We keep hearing about the brain drain; that there are too few scientists in all fields to meet our future needs; that the general public needs to know more about science, which is shaping his world to-day. My scientist friends and I feel that we have taken a step in the right direction in trying to strengthen the bond between the scientist and the educator. Together, we should be able to plan science courses to meet the real needs of a scientifically oriented society.

CHAPTER FIVE
THE CHALLENGE OF CHANGE—1967
A Panel on Industry Looks at Science in Our Schools

I was asked to participate in this teacher's conference to discuss the role of science in the industrial world. I had the impression that my message was rejected by most of those in the audience, judging by the questions and comments.

Since we are concerned with the challenge of change it would seem appropriate to examine briefly three aspects of science in our schools in the context of an industrial society, namely:

1. The compatibility of the aims of scientists and industrial employers.

2. The social consequences of science in industry; who is in the driver's seat?

3. The influence of the teacher in the making of a scientist and the development of his attitudes.

4. Then we must ask ourselves this additional question:

5. What are the necessary conditions in the teaching of science that arise from these considerations?

Are the aims of scientists and industrial employers compatible?

The obvious answer is both "yes" and "no". It seems to me, as a rank outsider, that the aims of industry are to provide for the mechanical needs of our society and this must be done in a way that is economically profitable. Also, it is evident that industry deliberately creates mechanical needs by convincing people that they need such and such a gadget or model change; in order to do this, industry makes use of scientifically derived technology. It is also true that industry is strongly influenced by national aspirations whether in war or in the conquest of space. Physical needs, such as roads needed to satisfy a highly mobile society, have stimulated the evolvement of huge earth moving machinery by means of which man changes his physical environment; to say nothing of the fouling of man's nest with pollutants.

In a different context, Stephen Cotgrove writing in the *New Scientist* pointed out that there is a conflict between the production manager in industry and the scientist who provides the ideas whereby he achieves his production. This conflict arises out of differences in aims and language and has a direct bearing on the creativity of the scientist.

I do not wish to go further into this point now, except to point out that the sociology of industry is very poorly understood and that industry cannot look at science in our schools solely as a means of obtaining people who will provide the input of ideas on which industry feeds in order to grow. Industry must be concerned with the conflict between the scientific mind and the management mind. I believe that, as teachers, you must be aware that your students may well be involved in this conflict. H. W. Brode of Bell Telephone Laboratories has pointed out that, historically, technology preceded science both in time and location until about 1850, and that the role played by the interaction of various disciplines and technologies implies that a broad knowledge is necessary in order to impart the significance of this interaction.

Now what about the social consequences of science in industry and how does this affect your role as science teachers?

Fred Hoyle in his book Of *Men and Galaxies* has pointed out very forcefully that for any civilization to survive on a planet such as the earth, the transition from a primitive existence to a technologically sophisticated existence has to be made rapidly, otherwise the transition will fail because of fuel exhaustion; we must be able to operate on nuclear fuels before the coal and oil reserves run out, otherwise we fail. It looks as though we have a fighting chance to survive. In our Western society, at least, we have evolved an inter-meshed industrial complex so intricate and vast that I am inclined to believe that no one is in control. I cannot imagine either governments or industrial management making decisions which are not predicated on the assumption that the physical machinery of our society must survive and that we as individuals benefit only incidentally. In the foreword to the *Proceedings of a Conference on Space, Science and Urban Life* held in Oakland, California in 1963, one of the objectives is listed as: "to consider and evaluate the application and relevance of new technology to the needs of industry and cities for new processes, products, materials and techniques which will enhance and stimulate the productive economy and industrial growth of the city, state and nation".

Seymor Melman in *Profits Without Production*, provides a remarkable insight into modern management practices that produce an increase in profitability, but also as a consequence, a large decrease in employment and productivity. It was concepts such as this as enunciated by Clark Kerr of Berkeley when he talked of the Multiversity that led to the Berkeley student riots a few years ago.

As teachers of science, how do you see your role in this situation? Do you have a responsibility to make your students aware that such a situation might exist or do you leave it to the history teacher, who may not even be aware of it? Remember, your students of today will probably be the ones who make the transition through the natural fuel exhaustion stage that Fred Hoyle talked about.

Now I come to the influence of the teacher on the making of a scientist—here we may be in more agreement! John Verhoogen has put it this way: "Human beings want bread, and they want freedom, and some of them want to know". You are dealing with those who want to know.

In reading for this conference, I came across several charts showing the influence of teachers of the subsequent flowering of a particular branch of science. Dr. Krebs, a Nobel Laureate in Chemistry, in discussing The *Making of a Scientist* in a recent issue of *Nature* wrote:

> "Above all, attitudes rather than knowledge are conveyed by the distinguished teacher. Technical skills can be learned from many teachers and like a modicum of intelligence, are, of course, prerequisites for successful research. What is critical is the use of skills, how to assess their potentialities and their limitations; how to improve, rejuvenate, to supplement them. But perhaps the most important element of attitude is humility because from it flows a self-critical mind and the continuous effort to learn and improve. Also of great importance is the enthusiasm conveyed from teacher to pupil".

All of us remember at least one teacher who stands out in our educational experience. It is almost certain that each one of these people had an inherent integrity that commanded respect, also whether consciously or not, they had a respect for the art of teaching.

Margaret Mead says that it is not teachers that our society looks down on, but teaching itself.

My point is simply this—that all the effort put into curriculum revision, new texts, laboratories, and so on, is of no avail if you as teachers fail to aspire to great teaching. This now leads me to my final comments. Great teaching can only come about in an atmosphere of intellectual and academic freedom and this freedom can only be obtained if you as teachers show evidence of professional dedication and competence. I believe that all teachers, no matter what their specialty, must have broad roots in our culture—they must read widely—they must be concerned with, and knowledgeable about, current ideas in education at all levels. An impossible task? Not really—a dedicated scientist works long hours—is often widely read in subjects outside his specialty, and has a keen interest in our society. Can you, who teach future scientists, be any less dedicated?

I would suggest that our present educational system almost totally precludes this type of dedication. We pay lip service to the ideal I have just described, but then we hedge in the teacher with boundaries and restrictions on curricula, texts, examinations, and above all, time.

Why?

Are you really not competent to handle the responsibility that academic freedom implies? Do the standards of admission set by universities in order to select their freshmen students demand conformity to which you must comply? Why then the common complaint of "unlearning" in the first year of university? Does the Department of Education not have sufficient confidence in their teacher training program? Is not the fostering of creativity and individualistic thinking important?

I would like to see educational authorities have the courage to say to their teachers: "We believe you are professionally competent—go and teach. Find out the demands of our society, our universities, our industry. Stand on your own feet. But above all, teach with integrity and respect for the students who pass through your classroom doors".

For, in spite of the challenge of change, these students, like all of us, are composed of a little bit of stardust and will be confined to the surface of this planet for the rest of their days. Must we not inspire them with the wonders of man's understanding of his universe and the dignity of being human?

CHAPTER SIX
THAT WHICH IS & MY CONVICTIONS—1967

This chapter emanates from an address to the Unitarian Congregation of Ottawa as part of their summer presentations by lay people. I began by describing some of my early thoughts about our existence and concluded with my current convictions.

That Which Is

This seems to be a summer of soul searching. As others have done in this series of talks, I will be going back to my younger days for the basis of my discussion this morning, to sketch in some of the anchor points in my attempt to understand reality.

But first let me clarify a point or two. My theme concerns some philosophical aspects of science. I have no formal learning in the field of philosophy, nor in the field of the science that I will be using to illustrate my points, namely, biology. This is, perhaps, useful since it puts us all in the same boat (except for any biologists or philosophers who may be present!). And so—metaphorically—let us push off from the shores that we know.

Imagine, if you will, a young lad during the mid-thirties on a prairie farm. The economic future is indeed bleak, but that is not what he thinks about. He has just finished his high schooling and is now aware of some literature, some history, geography, and also some science. He has also discovered philosophy through Will Durant's classic book, *The Story of Philosophy*, and he has had some religious education as well. It is about these ideas that he thinks as he rides up and down the fields on the plow or walks behind the harrow—and the days are never long.

Who is he? Why is he here, and by what means is he aware of himself? If all material things are made up of 92 elements (more were to be discovered later), how can such inanimate matter be living and aware of itself? Or is he just a figment of some cosmic imagination? Longfellow spoke truly when he said, "The thoughts of youth are long, long thoughts."

It is from those thoughts that the title *That Which Is* comes. After several years of thinking, the young lad, now in his twenties, decided that no matter what he thought about the reality of the world, or for that matter, what anyone else thought about it, it existed, and only the limitations of our knowledge of the laws of nature, prevented us from knowing that reality in all its cosmic majesty. It was from that time that he said to himself, "*That which is, is.*"

My Sunday reminiscences ended this way:

Truth is revealed to us through the immutability of natural laws. Man's quest for knowledge of the universe about him, and of himself, can only be revealed in terms that are consistent with those laws. No matter how we may misunderstand or be ignorant of the laws of nature they are valid and we, even in our ignorance, cannot violate them—no exercise of reason, logic or denial by our consciousness can in any way influence "*That which is.*"—we can only try to understand.

But what of joy and sorrow, love and hate, tragedy and triumph—are they not also real—can we not communicate with one another? How is it that I can say what I have just said and each of you receive some understanding of what I am trying to convey? Why do we respond to the elegance of a simple song of love?

I do not know; nor do I think anyone has the answer to these questions—but of this I am sure; these questions are about reality and that reality is related to the natural laws, though with a profound complexity. So I say go and enjoy the sunset— the velvety black of the star-studded mantle of the night sky—the beauty of music—the warmth of companionship—rejoice in the voices of your children—experience the horror and revulsion of war—the terrible ache of hunger, and, if you will, the omniscience of a Divine Being. These all have the reality of the experience of a conscious thinking being, but never suffer the delusion that nature will alter her ways because we do not understand her and that by sheer reason or logic we can construct a framework of belief that will stand against the simplicity, elegance, and beauty of an absolutely obdurate nature.

How can I say this with conviction? Again, I do not know—but I do know that all my experience, all my reading, all my thinking has not been able to undermine the simple statement "*That which is, is.*" This statement in no way illuminates the fundamental behaviour of nature or of man; this can only come through the long, frustrating, but endlessly fascinating searching that is the hallmark of mankind. What it has done for me, is to enable me to achieve an equanimity that has made many of the years of my life more meaningful to me than they otherwise might have been.

CHAPTER SEVEN
SCIENCE FOR NON-SCIENCE
ORIENTED STUDENTS—1968

Blessed is he who contemplates
the ageless order of immortal nature
how it is constituted, and when and why
Euripides

This is a statement of my position on a proposed high school course on Space Science and Astronomy.

For several years, I acted as a consultant for the Ontario Department of Education. I also devoted a week of my holidays to attending Science Summer Courses for Elementary School teachers where I talked about science and what it means to society. This statement was made without reference to the course details.

"In the past science acted on society through technology. Now society is reacting back on us. Do we amplify the cycle by intervening overtly? Or do we simply stand aside and allow things to go to the devil"? Astronomer Fred Hoyle put this question to the readers of *Physics Today* earlier this year. Variations of this question can be found almost every month in the current literature. In many of the articles are found a variety of remedies which are designed to combat the declining enrollment in science, particularly in physics.

Only recently has there been any evidence of a concern regarding the scientific literacy of the much larger number of students who are non-science oriented.

These students must not leave High School afraid of science or unconcerned about the relevance of the impact of science on our society.

The fact that this committee exists is a tacit answer to the question posed by Fred Hoyle. We cannot "stand aside and let things go to the devil". It is possible for this committee to devise an elaborate course structure—but to what end?

Recently, I attended a High School graduation ceremony during which the usual platitudes were uttered; then the valedictorian, with the brashness of youth, bluntly stated, to the applause of the assembled students that they were taught "what to think, not how to think". He commended those teachers who departed from the prepared syllabus and explored topics of interest to the students. It is significant that he had a summer and a month at university to reflect on his address.

Most of the applauding students were not science oriented if one could judge from the introductions as they received their diplomas. Yet, the technology that made the affluence of their lives possible, the public address system the valedictorian used to criticize the education he had just received, the glamorous synthetic fabrics adorning the mini-skirted graduates were all made possible by the essentially human ability to organize experience and to ask questions of nature—in a word, by science.

Herman Kahn and Anthony Wiener, in that rather depressing book "*The Year 2000*" say in the concluding chapter:

> "If there is any single lesson to be learned from the above, it seems to be this: while it would certainly be desirable and might even be helpful to have

a better grasp of how social action may lead to unanticipated or unwanted results, it is not likely to be sufficient. Given man's vastly increased power over his internal and external environment and, in particular, given the unprecedented opportunities for centralization of social control that follow from the economic and technological changes that have occurred and that are likely to with ever increasing impact, the effect of social policies, planned or haphazard, are likely to increase drastically and the consequences of mistakes are likely to grow correspondingly disastrous".

The operative words are "mistakes" and "disastrous". By implication, an intelligent, educated and scientifically alert public will have a much better chance of surviving.

What has all this to do with a science course on space and astronomy? The success, reported to this committee by teachers who have had the vision and ability to devise such courses, in stimulating, maintaining and expanding the interests of the academically weak students as well as those more strongly motivated, attests to the potential interest in such a course.

But why is interest in space science and astronomy "socially desirable" since those taking the course will be, almost by definition, not going to pursue a career in any branch of science?

To return to the trite platitudes of the commencement platform; by 1984 these students will be among those taking the social action that Kahn and Weiner consider may lead to unanticipated and unwanted results. By capturing their interest, by exciting their imagination, by acquainting them with a branch of science that at once is as new as tomorrow and has its roots as far back in antiquity as the questions man has asked of the sky above him, "What is up there?", "What and where are the stars?", "Who am I that I can ask these questions?"; by these means we can encourage them to ask questions of themselves that may lead to concerns that if science itself is amoral, and what society does via technology involves social decisions, what if anything is their role? Then some of the concerns and restlessness of modern youth may be directed from informed choice rather than outright rejection and protest; high school graduates may then become more literate scientifically than is now possible. In short, it may enable them to contend with the taboo against knowing who you are.

Does the course have to say all this? Certainly not. What is essential is that those who devise and advise, those who teach and those who administrate, recognize that a statement of objectives does not automatically ensure their accomplishment. The accomplishment will or will not occur in the minds of the students.

It seems to me that the prime objective is to provide a course content and climate such that an awareness will be stimulated in the student of the essential unity of his environment and that rejection by reason of ignorance, apathy or hostility of any part of it will leave him only partly educated and consequently more prone to make the mistakes Kahn is so concerned about. The student will, we hope have the successes that will enrich his life because of his heightened awareness of his total environment. Aesthetically, he will be able to appreciate the poet or folk singer, but he will also be aware that the effectiveness of the communication is enhanced by modern electronic means. Hopefully, he will also be aware that there is selection in such communications.

Not only do we need separately scientists, farmers, businessmen, and engineers as well as poets, but as Cecil King pointed out recently in the *New Scientist* "The great need of society, as I see it, is for our administrators to keep constantly in the front of their minds that it is essential to use scientific methods in dealing with governmental and business problems". He goes on to point out that, for too long, we have proceeded by "hunch" or on the "educated prejudice of the supposed expert". His plea is for more scientists on company boards of directors and in the higher Civil Service. That means he says that "we cannot separate the study of science and the study of society".

For a course such as the one we are proposing to succeed. it seems plain that the teachers involved have a glimpse of the background I have outlined. They should have some practical knowledge of how the scientific and technological communities function. Updating of teacher qualifications has traditionally been accomplished by taking summer courses either at university or teachers' colleges. If effective teaching in the context of this course implies experience in the ways of the world outside the teaching profession (the world the student will inhabit), then recognition should be given for practical experience, be it in factory, laboratory, business or travel. Recognition could perhaps be based upon presentation of a suitably researched and carefully prepared report on the educational relevance of such experience.

The impact of science on our society is increasing steadily. Successful implementation of a course such as this one will certainly give the non-science oriented student a much better chance of reacting intelligently to these changes rather than suffering the baffled alienations that Kahn and others are concerned about.

This, then, is my rationale for my participation on this committee and my hopes for its accomplishments.

(The course eventually became known as *MAN and SPACE*. It was not successful because it demanded more from the teachers than they were prepared to give. It also demanded a multi-disciplinary approach to teaching the course—again, the teachers were not prepared to operate in this way.)

CHAPTER EIGHT
SCIENCE, STUDENTS and SEMANTICS—1971

This paper was prepared for a presentation in the main auditorium of the Radio & Electrical Engineering building, M50, of the Montreal Road campus of the National Research Council. I subsequently used it in variously abridged forms for talks to educators across Canada. It is my distillation of the information obtained from the CAP (Canadian Association of Physicists) study of student attitudes toward science and technology.

Science

Science has been described as the logical outcome of the activity of the cerebral cortex of homo sapiens. He knows that he knows, and knowing this, he can order his knowledge. For millennia, this knowledge was regarded as sacred and was shrouded in myth and the trappings of authority and power. The gradual, but inevitable, evolution of the use of codifying symbols preserved and spread knowledge. A veritable explosion followed Gutenberg with the development of the printing press about 1450; we are all now aware of the ubiquitous power of the printed word. The use of the inductive method of reasoning, as suggested by Francis Bacon opened the flood gates, and now man has set foot on the moon. Clearly, science is part of man's value system. Yesterday, he perused science with fervour and reverence; but today, he is questioning that pursuit. Particularly, the young are questioning science and its value systems. They, as youth have always done, are evolving their own value systems and in that process they are forming their attitudes towards themselves, society, and inevitably, towards science.

At the risk of oversimplification, let me outline in capsule form how I think different segments of our society perceive science.

The Scientist

He seeks answers to questions related to the nature of the universe about him—often some minute portion of the universe. His motivation is curiosity and a deep intellectual satisfaction in just simply knowing. He is aware that his findings and understanding, i.e., his science can and often will be used by others. Most scientists regard scientific knowledge as value free.

Educators (Including teachers)

Science is obviously a significant part of human culture and as such should be part of everyone's education. Yet, at best, most introductions to science are second hand. The educator can intellectualize the experience of science, but rarely can he personally experience it, and so he teaches science more from necessity than from deep conviction. Always, he is teaching yesterday's science, while his students are living with the effects and implications of today's science.

The Public

The public equates science with technology, which is all about them, yet they view the scientist as an intellectual wizard who does mysterious things in remote laboratories. Also they, that is, the scientists, can always find answers to questions, e.g., *they* will find a cure for cancer. The public generally supports science because, somehow it is a "good thing," instead of an understanding of science and the role of scientific knowledge in our society. It must be noted that science, as a beneficial activity, is being increasingly questioned.

Politicians

Science and scientists are necessary and therefore must be supported; e.g., scientists are useful in times of emergency, such as war. In a way they are also a nuisance, since they always want large sums of money for large playthings that don't seem to have immediate practical value. Yet, it is increasingly obvious today that science is essential in the development nations, so politicians are developing "science policy." How they will fare in Canada we do not yet know.

Students

Science is part of the curriculum and so interferes with the enjoyment of one's high school experience. For the student who may be deeply interested in some aspect of science the rest of high school is a nuisance. Actually, the dichotomy is not nearly so clear cut, but girls probably regard science with more disfavour than do boys; physics is probably considered as the least desirable of all sciences. It is rarely regarded as part of our cultural heritage in the same way as literature or history.

Clearly, the stereotyped images I have just described represent a range of beliefs and attitudes regarding science and scientists. One can perhaps see a reflection of how students perceive science in a paper by Beardslee and O'Dowd, *The College Student's Image of the Scientist* (1961); some of his attributes are listed below:

- Highly intelligent; devoted to his work

- Individualistic; his loyalty is questioned

- Socially withdrawn; has few friends

- Indifferent to people; an unusual set of values

- Somewhat depressed; has difficulty controlling impulses

- Unhappy home life; his wife is not pretty

- Has an air of strangeness: plays chess, rarely bridge, never poker

- Influential but naïve: hard to like and comprehend

- Not interested in art; not colourful

- Uses his powers in a rational way

The late Dr. Hertzberg, Nobel Laureate, had a fine baritone voice and loved to sing opera. I knew many scientists who were and are artistic in many ways. Most of them frequented the National Arts Centre in Ottawa, where I was employed by the National Research Council.

I have built up an image of science and scientists with which you may not agree. I admit I have been selective. My purpose has been to show that such stereotyped images exist, and that they will show up in the measurements we made of the attitudes of high school students towards science and technology.

Students

The basic question to which we are addressing ourselves is: How can all youth grow and develop so that they can assume a constructive, humane management of our technologically-based economy, considering our increasing ecological awareness? Clearly, this management will require considerable scientific sophistication and rationality. Yet, if one can believe the media, there is a growing wave of anti-rational, anti-science sentiment developing throughout the world.

The flood of partial, unrelated and even pseudo information that pours over the inhabitants of the industrialized world cannot but produce an unrealistic vision of that world and its problems. Young people tend to emotional solutions of problems; but chronologically they are, by definition, immature and in our society impotent to achieve redress of the problems. Does this lead to an interest in the occult and astrology, as a way of seeking emotional release? Will some other agency, e.g., UFO's, bring solutions? Can answers be found by returning to nature and the commune? The hippie movement, which tends to "freak out" the adults by rejecting their standards of behaviour and values, can only survive in an indulgent, materially affluent society. To "relate" does not carry out the energy manipulation and transformations necessary to keep the political-industrial complex operating. The real and hard question is: how does mankind regulate his interaction with his physical environment so that both he and his environment have a chance of surviving?

The youth say we must change—the question is how, in what manner and at what rate? Man's ego-dominated arrogance towards nature is probably what youth rejects most fiercely. So now the question becomes: how can we ensure a transition from man's dominance over natural evolutionary processes, to an operative concern that both man and his world be cherished? In other words, can we as individuals mature without destroying the evolution of our physical environment? Are youth today maturing or are they being forced into perpetual adolescence? Let us, therefore, examine some of the influences on the young, particularly their schooling.

Here, one meets the system. If one recognizes it then one can survive relatively unscathed; otherwise one may never recover. Schools exist to facilitate the socializing process. Hence there are behaviour pattern norms to which the pupil must adjust. He must learn those behaviour patterns that his superiors, teachers, principals, administrators, parents, etc, decide are good for him. But he has learned other things from his peer group, his own experience, and these come increasingly into conflict.

Do the fantasies, myths etc. of childhood cause imprinting in early childhood experiences? How potent is this imprinting? Does his familiarity with the often unreal worlds of TV, comics, cartoons, intensify his humanity and his desire to be human instead of becoming a cog in the machinery of technology? Communications media comprise, among other things, TV, comic books, MAD, cartoons, books in general, newspapers, advertising, ownership of things, the occult, etc. Opinions, feelings, attitudes, data (factual, false, distorted) are communicated by these media. The child is exposed to this flood of information from his earliest days (or deprived, which is also a potent factor) yet the business of modern life restricts his time to mature and form his own view of the world. What is the filtering process that determines what gets through to the child? What happens as more and more experience and information overlays that obtained previously?

The ability to communicate non-verbally through images, mood music, "in" jargon, give youth a sense of solidarity, potency and authority. This is unfortunately not a substitute for wisdom, maturity and knowledge. Can teachers bridge this gap?

The young are aware of the *in thing* (cool) which is dominated by the group or the gang. He seems to be seeking his identity. He is still subject to the whims of society, be it paternal, legalistic or educational. But as a member of the group he is an equal among his peers. Affluence, mobility and changing family ties have given him time, money, the car, radio, TV, and the 'camcorder', so the group now becomes nation-wide. Youth have recognized their numbers and have learned how to communicate within their own culture. Cynics, particularly in the music industry, have exploited this ability to communicate. The result has been a developing awareness of power and that it can be used, if for no other purpose than to create action and excitement. Rock festivals are an exercise of that power for peaceful means, but in a different milieu from that of adult "square" society. Drug use also derives much of its iniquitousness from the nation-wide youth mobility.

A major problem is how to learn to listen to the anguish of youth through the noise of the adolescent peer group. Youth are saying that they want to choose what they shall learn, as in the tradition of the medieval university. The selection of available knowledge and skills and hence what is taught in the schools is determined by curricula. Curricula are devised by those in authority and, at least until recently, imposed on both teacher and taught. Curricula are limited by the filter network through which they reach the student, i.e. teachers, textbooks, etc.

The student is often unintentionally visualized as a black box having an input, output and a handle on the side. By feeding curricula into the input terminal and turning the crank, an output will appear known as the educated student. Unfortunately, the curricula is designed according to the output desired and little attention is paid to the nature of the input terminal, and virtually nothing is known of the real, as opposed to the imagined, contents of the black box. An impedance match is necessary for maximum knowledge transfer through the black box to the output terminal, the educated student. This requires measurement and knowledge of the box content, not in terms of its individual components, but rather in terms of their behaviour in response to certain stimuli. This impedance match is strongly conditioned by attitudes and this seems to be the neglected quantity when designing curricula having a specific output as an objective.

Recently Rokeach, writing in the *American Psychologist,* has shown that it is possible to induce relatively enduring behaviour changes in values, attitudes and behaviour as a result of brief experiences during periods of self-dissatisfaction. Changes induced during periods of time of only 30-40 minutes were still evident nearly two years later, as opposed to no effect on a control group. My point in mentioning this work is that student's attitudes may be particularly susceptible to long lasting change and modification during their adolescence when their general level of dissatisfaction and frustrations are high.

Semantics

We communicate formally through words; informally by body posture, facial expression, dress, etc. Nearly all the formal communication in the school system is through words. But words have meanings and those meanings may indeed be different for different individuals. C. D. Osgood, in his book *The Measurement of Meaning* described a means whereby he could measure meanings and he attempted to visualize the results of his measurements in what he called semantic space. He called his technique the Semantic Differential, and it has been used extensively since then in studying, among many other things, racial attitudes, marketing techniques and the development of meaning in school children.

During 1970, the Canadian Association of Physicists (CAP) instigated a study in universities and high schools of enrollment trends in the physical sciences. I was appointed the administrator of the Study Group. A major question was "What are the factors that are influencing these trends away from the physical sciences?" In the study, we made the assumption that students' perceptions of science and technology would influence their course selection and career choice. So, a measure of attitude would enable us to examine the hypothesis that an increasing percentage of students were rejecting the physical sciences, and that this rejection was related to societal influences. We hoped to be able to gain some insight into what the influences were and their relative potency.

This problem has been largely ignored by the social science and educational communities, so we were, in a sense, exploring virgin territory. One of our major problems was to find a suitable measuring instrument. The social scientists and the psychologists we consulted presented such a divergence of opinion (attitude!) that listening to all of them would have meant complete inaction.

The definition of attitude is difficult. It has proved to be almost impossible to predict behaviour from a measure of attitude only, since attitude is one of many variables that determine an individual's behaviour in any given situation. Recognizing this limitation and the need for a better definition of the phenomena that we were trying to examine, I will continue to use the word attitude, simply because it is so widely used in this context in the literature.

In this study, we assumed that a person's attitude toward an object would be a more likely determinant of his action, than his opinion about that object. In other words, his attitude will tend to determine what he actually does; his opinion is what he thinks he might do.

It is obvious that we are now deeply into the area of semantics—the meanings of words. It was in this area that Osgood made his contribution in providing a well defined technique that could be used in a controlled way. (We have since explored other methods, but these will not be discussed in this paper).

In Osgood's Semantic Differential technique the subject is asked to respond to a headline concept by marking a seven point scale between adjectival polar opposites, e.g. good to bad. We did not use the full potential of the technique and examine the independence of various factors in semantic space, but we did use it to get some idea of the semantic profile of various concepts and some idea of the ranking of the concepts.

Redundant scales and concepts were used and the presentation was randomized so that students could not deduce the basic hypothesis that was being examined. In other words, neither the students nor assisting staff were able to guess the purpose of the tests.

An additional consideration in using the S.D. was that Jenkins, et al., writing in the *American Journal of Psychology* used the S.D. in 1958 to prepare an atlas of the meaning of 360 words. Some of these words and scales were included in our study as a control. Subsequently, a study of the semantic attributes of 487 words as related to the conceptual behaviour of fifth-grade children was published in the *Journal of Educational Psychology* by F. J. Di Vesta in December 1970. But more on that later.

The results to be presented were obtained during developmental phases and do not represent a definitive technique. Three tests were carried out, the first two with a predominantly rural group of students and the second with an urban group of students. All three tests were different to some degree, but some common concepts and scales were maintained.

Normally, the scales are simply numbered 1 to 7 with no weights attached, however, the factor analysis using the data in this form required a larger computer than the one available to us at the time the data were reduced. An alternative method that required the imposition of a value system was used, in that we designated +ve and -ve values to the scales and weighted them as 0 for responses to position 4, and + or—1, 2, and 3 from the middle position. See Figures 1a and 1b.

Using this technique, the responses of each student to each concept were summed by adding the weighted positive and negative responses for the common scales used on all tests. For example, in one particular case the sum of a Grade XI boy's responses to TECHNOLOGY were +12; -12, while a Grade XI girl's responses to TECHNOLOGY were +14; -1, for the common scales. In this case, the boy's responses cancelled in the overall summing so that he would not be counted as having a negative response, yet he clearly had some strong -ve attitudes towards TECHNOLOGY. I point this out to show that our results are probably conservative.

The response of 79 urban students and 161 rural students is shown in Table 1. The balance by sex and grade is good. The ratio of the total +ve and -ve responses for all students to each concept was used to rank the importance of the concepts for the two student populations.

Table No. 2

79 Urban Students – Concept Ranking and Response
33-Male; 46-Female; 22-Grade IX; 28-Grade XI; 29-Grade XIII
161 Rural Students – Concept Ranking and Response
85-Male; 76-Female; 49-Grade IX; 67-Grade XI; 45-Grade XIII

DATA CONCEPT		MR*	MUSIC	FAMILY	HEALTH*	SISTER*	SCIENTIST*	SALARY	KITTENS*	TECHNOLOGY*	STUDY*	MAN	INDUSTRIALIST
+ve responses	URBAN	7.82	6.85	6.25	4.53	3.57	3.45	3.19	2.47	2.14	1.85	1.34	0.97
-ve responses	RURAL	9.20	-	-	12.0	3.74	4.68	-	3.83	4.56	3.72	-	-
% -ve responses	URBAN	4	4	9	15	17	17	16	17	25	25	41	49
responses	RURAL	2	-	-	2	9	8	-	4	8	5	-	-
% male -ve**	URBAN	6	9	12	18	21	21	12	12	33	21	58	46
responses	RURAL	2	-	-	2	12	8	-	6	7	5	-	-
% female -ve**	URBAN	2	0	7	9	11	11	15	17	13	28	28	52
responses	RURAL	1	-	-	1	5	8	-	1	8	5	-	-
% grade IX -ve	URBAN	5	0	5	5	14	23	9	14	23	23	40	41
responses	RURAL	0	-	-	0	4	4	-	6	4	0	-	-
% grade XI -ve	URBAN	4	7	14	7	18	11	21	11	12	25	36	52
responses	RURAL	0	-	-	3	11	9	-	2	6	6	-	-
% grade XIII -ve	URBAN	4	3	7	20	14	14	10	10	28	24	47	62
responses	RURAL	7	-	-	2	11	11	-	4	13	9	-	-

* These concepts are common to all three tests.
** Differences from %-ve responses above, are due to undecided responses (+ve= -ve) which ranged from 0-6%.

Responses from Rural students are generally more positive

Responses in the middle are regarded as neutral

TECHNOLOGY

TECHNOLOGY

slow	fast	slow	fast
cruel	kind	cruel	kind
helpful	harmful	helpful	harmful
masculine	feminine	masculine	feminine
untimely	timely	untimely	timely
active	passive	active	passive
mine	theirs	mine	theirs
unsuccessful	successful	unsuccessful	successful
hard	soft	hard	soft
wise	foolish	wise	foolish
new	old	new	old
good	bad	good	bad
weak	strong	weak	strong
important	unimportant	important	unimportant
frightening	reassuring	frightening	reassuring
calm	excitable	calm	excitable
false	true	false	true
colorless	colorful	colorless	colorful
unlikely	likely	unlikely	likely
honest	dishonest	honest	dishonest

Responses by an Urban student

Responses by a Rural Student

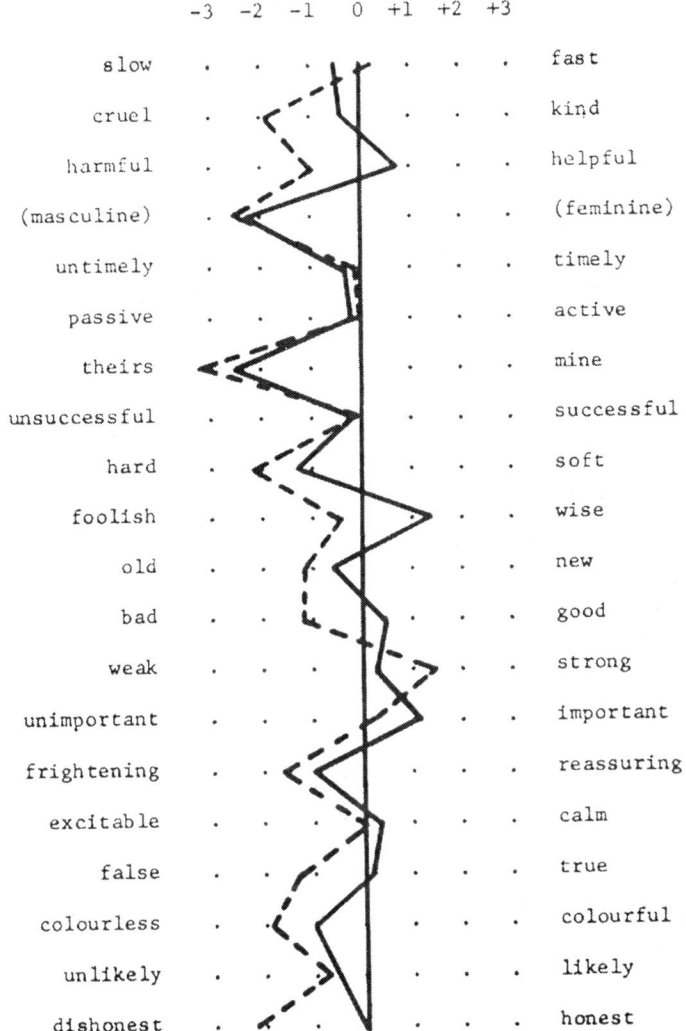

Figure 2: The concepts INDUSTRIALIST (dashed line) and SCIENTIST (solid line) normalized to the student self-image, i.e. the means of the responses to the concept ME (vertical line).

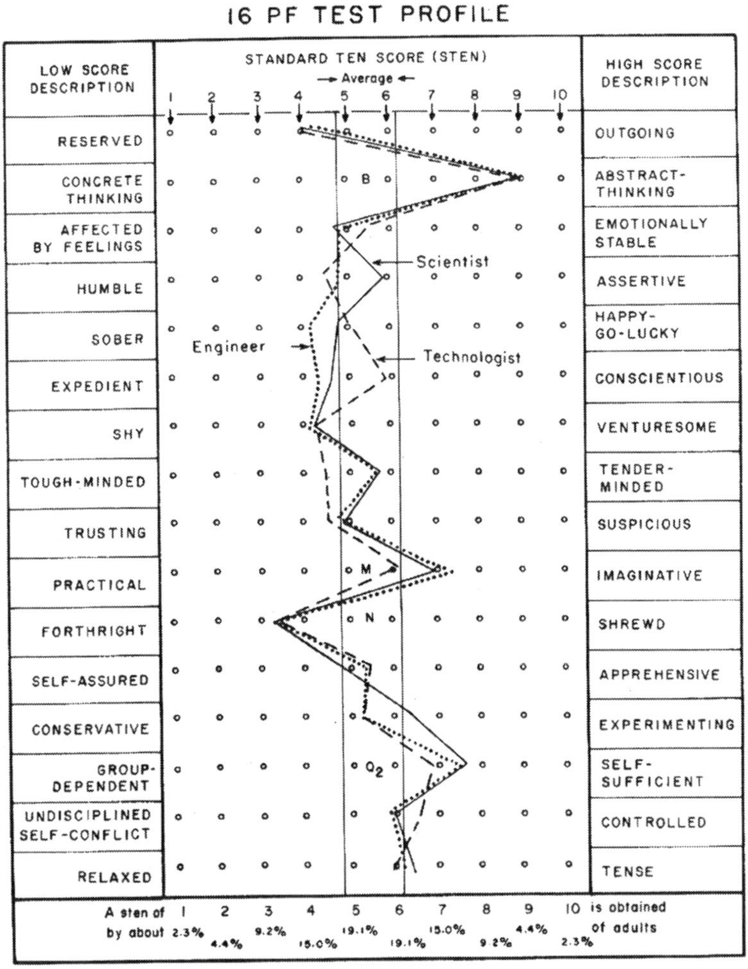

Figure 3: The measured personality profiles of 36 Scientists, 36 Engineers and 37 Technologists. The calculation of the Potential Scientist Pool is based on the four factors B, M, N and Q_2.

The following observations have emerged:

- The inversion of ME and HEALTH and the large difference in the ratio re HEALTH.

- The order of SCIENTIST, TECHNOLOGY, MAN and INDUSTRIALIST

- Difference in -ve responses between urban and rural.

Approximately 3:1 difference in -ve responses between boys and girls in the urban sample to TECHNOLOGY, compared with no difference in the rural groups. A similar difference can be seen in the response to SCIENTIST.

The rural student response by grade to SCIENTIST and TECHNOLOGY indicates an increasing -ve attitude as compared to a trend to more +ve attitudes towards SCIENTIST for the urban students, with little change towards TECHNOLOGY.

The strong -ve reaction to both MAN and INDUSTRIALIST was surprising, but perhaps understandable in the context of pollution and the "military-industrial-complex."

We thought it would be useful to examine the composite picture that the students had of SCIENTIST and INDUSTRIALIST. All students, both rural and urban, as well as those tested by Jenkins in 1958 have an almost identical self-image profile; their response to the concept ME is stable. Using the concept ME as a normalizing factor the students' images of SCIENTIST and INDUSTRIALIST are shown in Fig. 2.

The vertical line is the reference derived from ME and the other lines are plotted in terms of the difference of the concepts SCIENTIST and INDUSTRIALIST from ME. The scales were rearranged so that the somewhat arbitrarily chosen +ve and -ve values are on the right and left hand sides.

The INDUSTRIALIST is almost totally on the -ve side. Remember that these are the total responses of all urban students tested, not just those with negative attitudes. The scale mine-theirs can, I think, be regarded as an identification scale. Both the SCIENTIST and the INDUSTRIALIST seem to be strongly rejected. This is a composite illustration of how the average student views both the SCIENTIST and INDUSTRIALIST relative to himself.

Dr. Argyris, Chairman, Administrative Sciences at Yale University, in an article, *Students and Businessmen: The Bristling Dialogue*, says that:

> "... students attitudes toward the business world indicates the student thinks that: it doesn't care for the individual; it is primarily profit oriented; it is dishonest; it is routine; it is not creative and that businessmen are intellectually narrow."

The SCIENTIST fares somewhat better in that he is regarded as helpful, wise, good and important. But he is also hard, old, frightening and colourless. (Check the attributes of the scientist as found by Beardslee and O'Dowd above.)

Now the question is: how did the image of the scientist get this way? While we did not examine the concept SCIENCE per se, it is not unreasonable to assume that some of the attitudes toward scientists are projected onto the concept of science itself.

The words used in the teaching of science in the elementary schools have a positive and uplifting connotation. Since science is taught as part of the curriculum it automatically must be "good". The textbook says "a good scientist does so and so." The teacher presumably has a positive attitude toward science as a subject. So where does the attitude that the scientist is frightening come from?

Earlier, I mentioned some of the work done by Di Vesta on conceptual behaviour by fifth grade children. On a friendly-unfriendly scale of 1 to 7, ME, MAN and FAMILY were rated at about 2, i.e. strongly biased toward friendly, while SCIENCE was rated as 5.6 or unfriendly. The author makes a significant observation in discussing the results:

> "... It can be observed that a general impression or affective tendency toward an attitude-object, remains long after the specific knowledge (words) associated with the attitude-object, even though learned at one time, are forgotten." In another article on The Semantic Structures of Children? Di Vesta says "The mode of experiencing the environment and the way in which the experiences are encoded, with regard to the development of connotative meaning, appear to be securely fashioned by the time the child is in the second grade."

Rokeach, in a recent paper entitled *Long-Range Experimental Modification of Values, Attitudes and Behaviour* described how he was able to induce long lasting changes in values during an experimental involvement period of only 30-40 minutes, yet the results were measurable nearly 2 years later. He asks several relevant questions in his conclusions:

> "To what extent should our educational institutions shape values as well as impart knowledge, and, if so, which values and in what direction? If we have indeed learned how to bring about changes in values, attitudes and behaviour, as I think the experiments described here suggest, we must make certain that this kind of knowledge will be put to use for the benefit rather than the detriment of mankind."

What we have indicated so far, is that many students have measurable negative attitudes toward science and technology; and that these attitudes can be long lasting and may have their beginnings early in life. Is it possible to teach science to students who have a feeling of antipathy toward science? Are the early formative years the source of behaviour patterns in later years?

Celia Stendler writing in the *American Journal of Physics*, supports the idea that successful physics students are children who:

> "have in their early years been exposed to experiences that have built into their nervous system those models or constructs of the environment that facilitate the learning of the physical sciences; it is the thesis of her paper that more students could master high school physics if such experiences were a part of science curriculums in the elementary schools."

It may be possible that it is a deep-seated attitude due to feelings of success, that is operative rather than neural changes as Dr. Stendler implies. As Di Vesta has indicated, an affective tendency toward an attitude-object remains long after the specific associated knowledge is forgotten.

Another indicator is given by W. W. Cooley in a discussion of *The Potential Scientist Pool* where he points out that students opt into the potential scientist pool up to grade 11-12 and opt out after these grades and that these decisions are not related particularly to ability, but are related to attitudes and the expectations of significant others, i.e. parents. He says that the beginnings of an interest in science can often be traced to elementary school years.

This now leads me to report on some tentative findings regarding the attitudes of elementary school teachers. I had the opportunity, this summer, to give the same Semantic Differential Test to a group of about 40 K-Gr8 teachers, who were attending a Science Summer Course.

Since many of the teachers taught in small towns or rural areas, I used most of the concepts given to the rural high school students and added MATHEMATICS, to test the popular conception that such teachers personally dislike mathematics and communicate this dislike to their students.

The teachers were randomly selected from those attending the showing of a movie about Astronomy.

For comparison, the student's -ve responses ratio are shown in column 2.

The ranking is interesting, particularly the high ratio given to FAMILY. These were mainly married women teachers and it is probable that their experience has placed FAMILY high in their hierarchy of values. It was surprising to me to find that they put a somewhat lower value on HEALTH than did the students. The sample was too small for accurate results. The similarity with test results for students is striking with one major exception; the students seemed to understand the difference between ASTROLOGY and ASTRONOMY better than did the teachers!! They both rated HOROSCOPE last.

The teachers placed MATHEMATICS right below MUSIC. The popular conception that mathematics is disliked certainly is not supported by this listing of concept rankings; their ranking for mathematics might be influenced by an appreciation, as teachers, of its educational value, not their own valuation.

ENGINE and TECHNOLOGY were given an identical ranking, which is reasonable. However, the relative rankings of SCIENTIST, TECHNOLOGY, MAN and INDUSTRIALIST was the same as that of high school students and the ratios were of the same order of magnitude. So, one can infer that they have similar feelings toward these aspects of our society, except that there were no totally -ve responses to TECHNOLOGY.

What really surprised me was the high ranking of ASTROLOGY as opposed to HOROSCOPE. The only completely -ve responses to ASTROLOGY were given by teachers in the higher grade levels. Perhaps the effect of showing a film on astronomy had an influence, but I should think the teachers would appreciate the difference. The students seemed to be more aware of the difference.

TELEVISION was ranked lowest (apart from HOROSCOPE); 21 per cent responded negatively to TV. To discuss the effects of TV would require a paper in itself. There is no doubt that much of the negative attitude toward science is derived from TV and other media. These teachers certainly do not look on it favorably, relative to the other concepts presented to them. (See Chapter Twenty-Two)

There were five teachers who refused to identify themselves by Grade. These teachers had strong negative reactions to most scales except FAMILY, ME and HEALTH. They probably reacted negatively to the whole test concept, since they were not told what the test was about—that would have destroyed its validity.

The most significant impressions are:

- The teachers in the Junior grades are positive, outgoing people, and this is good.

- The incidence of -ve responses increases with grade level; this is worrisome.

These findings are from a limited, biased sample (they were motivated to take a science summer course) and should not be taken as representative of elementary school teachers as a whole. However, as with the tests of high school students, they do show trends that are not inconsistent with other data.

The real significance of this type of test is that it is simple to administer, complicated to process and can be easily repeated at later times to measure change. If such tests could be further developed and validated and a sampling of diverse populations made, we might have a much better picture of what is happening in education, and, in particular, we might be better prepared to meet the challenge of change.

It is axiomatic that the older members of any society regard teaching the young as one of their major obligations. In most societies this function is formalized in the institution of the school. A major category in this function is the teaching of science. If we use a dictionary definition of science as a branch of knowledge or study dealing with a body of facts or truths systematically arranged and showing the operation of general laws, then the teaching of science involves an introduction to the technique of systematizing knowledge and the operation of general laws.

But is it necessary to start at square one? Even the kindergarten child is immersed in a society that is systematized. What else does Sesame Street do? There is the classic story of the two five-year-olds discussing the jet aircraft flying overhead as the recess bell rings; one turns to the other and says "Aw, come on we gotta go string them beads!"

This sophistication is not knowledge, but can and does lead to dissatisfaction with curricula when students enter their teens and are confronted with the school system in the context of their developing awareness of the problems of mankind; they tend to question the relevance of the system.

The results presented here tend to confirm the picture of mankind as held by the student; he regards mankind as cruel, harmful, foolish, frightening and dishonest. One is not certain what is meant by this, nor whom they are evaluating—the "over thirties"? They apparently regard the scientist as generally helpful, but frightening and the use of the fruits of his labour, i.e. technology, as used by the industrialist as detrimental. It is disturbing, but not surprising that some teachers seem to have the same attitudes.

To me, this seems to be anti-rational. Remember the opening definition of science as the logical outcome of the activity of the cerebral cortex of homo sapiens. Modern research is tending to confirm this statement yet the behaviour of the youth of the species, homo sapiens, tends to refute it; hundreds of papers, speeches, etc. in recent years have deplored the anti-science, anti-rational outlook of youth all over the world.

A. B. Hodgetts's study, *What Heritage?, What Culture?*, has resulted in the formation of the Canadian Studies Foundation headed by the Honourable Walter Gordon. The object of this activity is to improve the cultural sensitivity and knowledge of Canadian children about their own country. Surely, it is equally important to have knowledge of the physical environment and the nature of the laws that govern man's interaction with that environment.

How can we solve the problems of pollution from an emotional rejection based on ignorance? It requires intellectual rigour to recognize the significance of rate of change. Emotionally and intuitively, people seem to extrapolate their past experience linearly into the future. Consequently, they neither see the past in perspective, nor do they understand the headlong rush into the future.

Let me digress slightly to examine a possible source of a difficulty directly related to teaching. McKinnon and Renner in the September issue of the *American Journal of Physics* (1971), ask some significant questions and, at the risk of mutilating their argument, I will quote some of the significant portions of their paper. Is the unrest today in many universities caused by student evaluation of problems based upon emotion rather than logic?

Do student claims that curricula are irrelevant, trivial, and inadequate in terms of the magnitude of the problems facing mankind today have substance, or are these students unable to evaluate logically the structure and necessity of these curricula? These questions, together with suspicions voiced by various professors of science about the inability of their freshman students to think logically about the simplest kind of problem, led the authors to question whether or not most college freshmen do think logically. This doubt about the ability of the entering freshmen to think logically led to the following hypothesis: *The majority of entering college freshmen do not come to college with adequate skills to argue logically about the importance of a given principle when the context in which it is used is slightly altered.*

The findings are that 50 per cent of the entering college students were operating completely at the concrete level of thought, as defined by Jean Piaget, and another 25 per cent had not fully attained the established criteria for formal thought. The males scored significantly more at the concrete level than the females.

McKinnon and Renner go on to assess the probable reasons for this situation. In summary, they do not find the major fault with the high schools. Is it then the elementary schools? They examine Piaget's criteria that intellectual development is dependent, among other things, on social transmission, pointing out that most often only information is being transmitted by the instructor. They then say,

> "Therefore, the blame must, in the last analysis
> be placed, at least partially, upon the shoulders of
> those who teach at the college level and who insist
> upon ignoring the rapidly accumulating evidence
> in favour of the enquiry approach."

The total accumulation of research to date leads to the following hypotheses:

- The secondary educational experience does not now promote logical thinking in most students.

- An abundance of enquiry-oriented courses, taught by teachers who are products of college and university professors who practice and profess enquiry, must come into being in the secondary schools before an alternative to the first hypothesis can be accepted. Those experiences will have to be developed by many colleges.

These hypotheses have profound educational implications since a serious problem has been shown to exist and the means for its alleviation have also been shown to be available to the profession. If colleges and universities do not try to solve the problem by assuming the responsibility for the intellectual development of their students, but continue to look at their primary purpose as the transmission of information about the several disciplines, the elementary and secondary schools will continue to fail in their mission of truly educating students. The needed changes, however, can only come through acceptance of an attitude of inquiry by all of those who teach the teachers.

The implications of the upgrading of teacher education to university level are profound. If the adage that "teachers teach as they are taught" has any validity and McKinnon and Renner's hypothesis is substantiated, then the universities, rather than being the cutting edge of man's probing into the future, may well be the millstone that holds him in the past while the mystical future, like the Pied Piper of Hamelin, calls the young to follow, all unprepared, into a country for which they cannot read the road map.

Mankind is irrevocably committed to a life increasingly dominated by science and technology; he has to make the transition from the use of natural fuels for his energy sources to technologically derived fuels (nuclear and/or solar energy), if he is to survive on this planet— and this survival also depends on the recognition that bigger, newer and faster are not synonymous with better.

Anti-science, anti-rational attitudes will only lessen the chances he has of making this transition. Man is also an aesthetic creature who has sensitivities, hope, joys and fears; he engages in anti-social behaviours such as war and the arbitrary exercise of power. How will he fare in the future? The answer depends in large part on how we teach the young.

CHAPTER NINE
OF PHYSICS and MEN—1972

This was published as an invited editorial in *Chem 13 News*, No. 40, p 2, May, 1972 by the Chemistry Department, University of Waterloo, Ontario. I had presented my talk on *Science, Students & Semantics* at the University, and subsequently was asked to write an editorial for the *Chem 13 News*.

In the sense that physics is the exploration and elucidation of the behaviour of fundamental particles, which are the basic constituents of all matter, it can truly be said to be a fundamental science. In turn, it would appear that the whole of physics finally rests on a few conservation laws which govern the interactions of elementary particles. Chemistry too, is concerned with matter and energy at a more complex level, in which particles in association have their individual behaviours modified to produce more complex structures, which in turn have different properties. The whole of what we now call the biosphere can be said to be the result of the chemistry of nature. Yet the permissiveness of nature is not unbounded.

The bounds are contained in the laws of physics. It would seem that in nature anything that can happen may happen, and that the laws of physics are primarily concerned with our attempt to understand what nature does not allow to happen. The structure of nature does not depend upon our understanding. Our understanding, however, has allowed us to assemble the constituents of nature (for us at least) in new ways, to create new elements and perhaps momentarily, new particles. Obviously, physics does not depend for its validity upon the understanding of every living human being. Yet, each of us is inherently subject to the manner in which the universe functions. Science is the continuing attempt to elucidate this behaviour and so make it comprehensible.

The use of the word 'comprehensible' immediately puts this discussion on a different plane. Each of us is made up of a combination (to use the word loosely) of fundamental particles. Biology is the study of such assemblages that are self-replicating or living. It is the living organism called man who has the necessary self-awareness which enables him to ask questions in the disciplined way called science, and then attempt to comprehend the universe in all its complexity.

That comprehension is leading to the belief that man will come asymptotically close to the explanations of his own awareness using the disciplines of science. In principle, this appears to be possible, in practice, the hierarchy of complexity seems to increase with our understanding of the inter-relatedness of the natural world. As we gain insight into the electronic and chemical behaviour of the constituents of our own bodies, so we also are gaining insights into the complexity of the association of men we call society. It is not the physical laws that make such assemblies complex, but rather the organization of the assemblies; individual ants seem to function aimlessly, yet a colony of hundreds of thousands of ants may show a discipline and organization that borders on intelligence. Only indirectly can we say that physics is basic to the other sciences. The challenge to comprehension and understanding lies in seeing the inter-relatedness of the natural universe, as opposed to the models of the universe that each of us now comprehends. It is this challenge to comprehend that man faces.

Each man has his own awareness and that awareness has been expressed in his art, his religion and his science. Yet, it would seem that uncomprehending nature, within the limits of those basic physical laws will allow us to do anything.

Our challenge, be we physicists, chemists, biologists, engineers, or indeed poets, is to try to comprehend and then to act within that human constraint we call compassion. The scientist and the poet are first of all human, with all that implies, secondly they both are subject to the obdurate universe and finally they are both trying to comprehend it. Let us therefore listen to each other and try to understand what each is saying.

CHAPTER TEN
LEARNING PROCESSES AND THE COMPUTER—1973

This paper was prepared for presentation at the CAP Congress at the University of Montreal in June 1973. I was a member of the CAP Science Education Study Committee; I thought that it might be useful to examine the question of learning, in view of the increasing usage of the computer in learning processes. My presentation was severely criticized at that time, but subsequent research has vindicated much of what I said.

Benjamin Bloom in his "Taxonomy of Educational Objectives" identifies three major descriptions of student behaviour. These domains are: the cognitive, which deals with the recall or recognition of knowledge and the development of intellectual abilities and skills; the affective, which deals with changes in interest, attitudes and values, and the development of appreciations and adequate adjustment; and finally, the manipulative or motor-skill area.

He points out that only the cognitive domain could be adequately dealt with at the time of writing (1956) and I expect that still holds today. Almost all of the literature dealing with pedagogy is concerned with the cognitive domain. This comes about because teachers are not at all clear about the learning experiences that are appropriate to the affective domain. Bloom also says that so little is done specifically in the motor skill area in secondary schools and colleges that it was not even possible to consider this domain in his taxonomy.

We all recognize the importance of motivations and attitudes in the learning process, yet the psychologists have not yet shown teachers how to measure achievement in this area. B.F. Skinner, of course, would say that the gratification of self-satisfaction is the operating motivation in his "operant conditioning" technique. Learning is a process that takes place deep in the mind (while most of us know what we mean by words such as "mind", we cannot describe it; the use of many such terms will have to be accepted in this brief discussion, simply because we have no alternative). We can all recall the experience of trying to understand some particular problem and then after the understanding has come, we wonder why it was so difficult, but we cannot say just what happened to enable us to understand. (*Current neuroscience is beginning to make this process comprehensible.*)

Learning seems to be a way of internalizing experience and action; speech seems to be one way of externalizing experience. Speech, of course, involves the use of language, which in turn is a symbolic expression of an internalized interaction with the sensory inputs from an external world or "reality." We cannot, each of us, experience the "reality" of another, hence to communicate about it we "symbolize" it. In order to preserve this experience we again symbolize it in the form of writing. Speech and writing are not synonymous. Speech seems to be universally present and may be innate, or at least it is very easily learned, while writing and reading have to be learned, often with some difficulty.

Noam Chomsky has made a strong case for a universal grammar that is somehow innate and that our use of language is possible because all of us have a deep innate structure that contains the necessary and sufficient conditions which allow a meaningful symbolization of reality. The surface structure is the particular language symbolization (word-object relationships and grammatical- variations in sentence structure) that link the deep structure to the outside world. For example, as does English, which we are now using.Noam Chomsky reminds us of a very interesting and rather neglected lecture given by Charles Sanders Peirce more than fifty years ago, in which he developed some rather similar notions about the acquisition of knowledge in general. Peirce argued that the general limits of human intelligence are much more narrow than might be suggested by romantic assumptions about the limitless perfectibility of man. He held that innate limitations on admissible hypotheses are a precondition for successful theory construction, and that the "guessing instinct" that provides hypotheses makes use of inductive procedures only for "corrective action".

Peirce maintained in this lecture that the history of early science shows that something approximating a correct theory was discovered with remarkable ease and rapidity, on the basis of highly inadequate data, as soon as certain problems were faced; he noted "how few were the guesses that men of surpassing genius had to make before they rightly guessed the laws of nature." And, he asked,

> "How was it that it happened by chance, because the chances are so overwhelmingly against the single true theory in the twenty or thirty thousand years which man has been a thinking animal, ever having come into any man's head."

A fortiori, the chances are even more overwhelmingly against the true theory of each language ever having come into the head of every four year-old child. Continuing with Peirce: "Man's mind has a natural adaptation to imagining correct theories of some kinds … If man had not the gift of a mind adapted to his requirements, he could not have acquired any knowledge." Correspondingly, in our present case, it seems that knowledge of a language—a grammar— can be acquired by an organism that is "preset" with a severe restriction on the form of the grammar. This innate restriction is a precondition, in the Kantian sense, for linguistic experience, and it appears to be the critical factor in determining the course and result of language learning.

The child cannot know at birth which language he is to learn, but he must know that its grammar must be of a predetermined form that excludes many imaginable languages. Having rejected a permissible hypothesis, he can use inductive evidence for corrective action, confirming or not confirming his choice. Once the hypothesis is sufficiently well confirmed, the child knows the language defined by this hypothesis; consequently, his knowledge extends enormously beyond his experience. Chomsky is convinced that if we are to understand the mind and mental processes, i.e. learning, we will in large part succeed by studying the deep nature of language.

Samuel Reiss, in his book "*The Rise of Words and Their Meanings*" has pointed out "the meaning of a sentence is to be inferred, not from a precise grammatical analysis, but from the sentence as a whole and from the context in which it occurs."

Reiss argues:

> "The simple common words of the native vocabulary of any language constitute one large family, that is an organic whole." He then goes on to show that "Every common word denoting explicitly a striking action of some kind implicitly involves in its meaning the sound, given by the word, that is associated with or is suggestive of the indicated action." Next he shows that all simple English words are "striking" which he further extends to "all simple non-English words are striking."

He does this with a number of non-related languages such as English, Hungarian, and Chinese. Of Chinese he says that "the entire basic stock of words constitute a single phonetic-semantically inter-linked family, that is an organic whole." He points out that in his view, the common striking kinship (action describing kinship) must be superimposed on any generic kinship that languages possess. In his view then, it is clear that in the deep underlying meaning of words, mankind does indeed speak one language.

It seems to me, therefore, that if we are to make any real progress in understanding the learning process from a pedagogical point of view, much attention will have to be paid to the ideas of such men as Chomsky and Reiss.

Let me now turn to a consideration of learning from the aspect of physiology. What happens to us as organisms when we learn? Dr. Celid Lavotelli (née Stenaler) in an article *Cognitive Development in Children and Readiness in High School Physics* advances the hypothesis that superior physics students, in addition to being bright, have in their early years been exposed to experiences that have built into their nervous systems those models or constructs of the environment that facilitate the learning of the physical sciences. Later in her article she says that whether a particular event serves as a stimulus and what the impact of that stimulus will be, depends upon the configurations already present in the cortex (of the brain). The cortex, in computer language, with its large storage capacity, becomes an information processing system holding complex strategies or programs, the context of which has already been determined by the previous experience of the system. (In modern terms: the unconscious.)

Learning can not only be thought of in terms of the computer analogy, but is also affected by the chemistry of the brain. For example, B. W. Agranoff has shown that if a gold fish is trained to perform a simple task and shortly thereafter a substance that blocks the manufacture of protein is injected into the skull, it forgets what it was taught.

M. R. Rosenweig et al, have measured the changes in the ratio of the weight of the cortex, to the weight of the rest of the brain for rats that have had enriched experience relative to control rats that had no enriching experience. They did indeed find that such enriching experience increases the weight of the cortex. It would appear that not only was the area of the individual synaptic connections increased, but also that the number of spines on neural dendrites was increased, i.e. the potential neural contacts and the area of those contacts was increased for those rats that had enriched experience. They speculate on the significance that an analysis of the effect of experience on the human brain might have if suitable experiments could be devised.

Sixty-five per cent of human brains examined at the Harvard Medical School showed greater development in the auditory area of the left hemisphere, compared with the same area of the right hemisphere. It is known that 90 per cent of people have left hemisphere speech dominance. Rosenweig then wonders about the effect on the right hemisphere in the case of musicians and concludes that we may be able to look for anatomical and chemical changes in the human brain that are correlated with experience and learning.

Erick R. Kandel, in discussing nerve cells and behaviour, indicates that habituation and dis-habituation appear to involve a change in the fundamental effectiveness of previously existing excitatory connections and that the habituation and dis-habituation are separate processes that occur at the same synapse. He says finally, these studies strengthen the assumption that a prerequisite for studying behavioural modification is understanding the neural circuit underlying behaviour. The same strategy may be applicable to biochemical studies of learning that will eventually describe its mechanisms at a molecular level. Kandel worked with a species of snail in which it was possible to identify specific neural cells and observe specific behaviour patterns as a result of controlled stimuli. The significance of such work on the larger question of learning is important. (Eric Kandel's recent books; *In Search of Memory* and *The Age of Insight* bring this fascinating topic up-to-date; 2012)

Let us now consider the question of the chemical transfer of learning. Can we modify the synaptic neural process and so alter habituation as Kandel calls it?

Many of us are aware of the famous flatworm experiment of a few years ago. Currently, more sophisticated experiments have been conducted with rats. The results of these experiments are the subject of a rather heated controversy in the literature and so must be regarded with some skepticism. In essence, rats were trained to perform certain actions. RNA or other identifiable substances not found in the brains of untrained rats was extracted from the brains of both trained and untrained rats from the same genetic stock. This substance was then injected into the brains of naive rats. Those injected with material from the brains of trained rats learned the original task statistically easier and faster than those rats who received material from the brains of untrained rats. A good deal of the controversy is centred on experiments in which a unique chemical was extracted from the brains of rats, trained to avoid the dark. This chemical was synthesized and injected into naive rats that, it is claimed, then showed immediately a statistically significant avoidance of the dark.

In humans it has been found that the amount of RNA in the brain increases, on the average, until about 40 years of age (life begins at 40!) and then remains stable until about sixty, when it declines sharply. Experiments on senile humans using RNA yeast (high in RNA content) have apparently increased alertness and improved short-term memory.

Learning implies retention of information, i.e. memory. Most of us are aware that we have a short-term memory. You can remember the words just spoken, but not the exact words said 15 minutes ago (although you can, I hope, remember some of the content of those words!). We also have a long-term memory. It would appear that the short-term memory is a temporary, working memory that is operative for less than 30 seconds, unless a rehearsal process is employed to update or "re-write the trace." This rehearsal is essential for transfer of information into the long-term store. The work of Wilder Penfield has shown how enduring that store can be. It also appears that forgetting is a failure of memory search during the retrieval process, rather than a degeneration of the memory itself.

Memory is selective in the sense that objects are more easily recalled than pictures which, in turn, are more easily recalled than words. Adults perform better than children, except that the recall of objects is very similar. One is tempted to speculate on the significance of this order in an evolutionary sense.

Recently, some interesting experiments have shown that an experienced typist can type at a slightly slower rate, with negligible change in error rate, while reciting well known nursery rhymes! This was also possible when she was repeating aloud from an audio input via headphones, while typing, provided that the two inputs were quite distinct in content. When the two inputs overlapped, neither task could be performed independently. Whether this is an example of two separately functioning systems, or some form of time-sharing is not clear, but it does show that it is possible to attend to two tasks simultaneously.

Earlier, the dominance of the left hemisphere in the language role was mentioned. D. Kimura has elegantly shown that the right hemisphere plays a dominant role in man's perception of his environment. Interestingly, tactile perception of Braille by blind people is more rapid with the left hand (right hemisphere) than with the right hand.

The right hemisphere is more effective in processing tactile information inputs. The sensory and motor systems are crossed to a greater or lesser extent. Thus the left hemisphere controls the motor functions of the right side of our bodies. Each hemisphere of the brain is also directly connected inside the skull by a network of nerves called the Corpus Callosum. Severing the corpus callosum has led to some interesting and significant experiments which, unfortunately, time does not allow discussion in this paper. Kimura has also shown that the right hemisphere specialization is more pronounced in males than in females; for example, males can more easily identify a design after rotation of the design than could females. On the other hand, females tend to have greater verbal fluency than males. She also speculates on the evolutionary significance of the difference in the intellectual organization of the male and female brains.

Recently, it has been reported by H. Landsell and J. C. Davie that another cross connection in the brain may be significant. A narrow band of nerve fibers connecting the two halves of the primate thalamus, under the cerebral hemispheres and below the corpus callosum, is missing in one-third of the male population and in one-quarter of the female population. When this connection is missing, that person performs better on non-verbal tests than do those who have the connection. This difference applies only in the case of men. It is speculated that the lack of this connection reduces the "cross-talk" between the two hemispheres and thus allows more acute attention to the spatial environment.

One final comment before we leave the brain. The brain has often been compared with the computer. In order for the computer to function, it must have an initial factory wired-in program, which is subsequently augmented by programming instructions. The outside world establishes contact with the computer through its monitor program (DOS in our terms today). If the monitor has to be changed or modified, the computer is not available, and such a change may be regarded as an internal "off-line" modification. This point, while somewhat simple-minded, is raised since sleep has been discussed in such terms.

M. Jouvet has described a state of deep sleep as "paradoxical sleep", since the cortex brain waves show all the signs of alert wakeful activity. This sleep state is also known as REM (rapid eye movement) sleep. The percentage of time spent in deep sleep increases with increasing evolutionary complexity. The cat spends most of the time in REM sleep, about 60 per cent of its sleep compared with 20 per cent for man. The young of all animals that experience REM sleep spend more than twice as much time in deep sleep as do the adults. Jouvet is quoted in a recent article as stating that he believes that during deep sleep a genetic activity takes place that re-creates us. This is the activity that has been likened to the off-line re-programming of a computer (a dubious analogy). It would, nevertheless, be interesting to examine the relationship between REM sleep and learning activity. M. J. Fowler et al., have conducted experiments that confirm the suspicions of many psychologists that sleep and memory are related.

E. H. Lennenberg insists that the development of language in children can best be understood in the context of developmental biology. He points out that the language development begins when the physical indications of brain maturity (e.g. gross weight, neuro density in the cerebral cortex) have reached at least 65 per cent of their mature value. It is interesting to note that by this time the degree of REM sleep has also decreased markedly.

Brain maturity in Lennenberg's terms is achieved at about 16 years of age. He postulates that, during development, the organism comes to a series of cross-roads; if condition A exists, it goes one way; if condition B is present, it goes another. Some of these conditions will be genetic, and some will be environmental.

Lennenberg concludes that man's capacity for language is species-specific, not because man is unique, but rather, because every animal species must be assumed to have specific cognitive attributes. In his view, "the brain is a biochemical machine"; it computes the relationships expressed in sentences and their components. It has a print-out mechanism consisting of acoustic patterns that are capable of similar relational computations by machines of the same constitution using the same program.

" ... Traditionally, learning theory has been involved neither in a specific description of this particular machine's behaviour nor its physical constitution. Its concern has been with the use of the machine: What makes it go? Can one make it operate more or less often? What purpose does it serve?"

Human learning is strongly language dependent. Chomsky maintains that our use of language is governed by a universal deep structure that is innate. Reiss maintains that the words we use have a universally similar origin that derives from the use of sounds that imitate simple "striking" sound producing actions. Since man has been evolving for some millions of years, it may not be surprising to find Chomsky's deep structure and Reiss' word origins embedded in our genetic make-up.

It is also interesting to speculate on Pierce's idea that there are innate limitations on admissible hypothesis as a pre-condition for successful theory construction. Could it be that we cannot, by our very nature, construct hypotheses that violate nature? I have often wondered how it is that pure mathematicians could construct systems of thought that later turned out to be so elegantly useful in physics.

Man is also a conscious, feeling animal and no matter how complex the relationship of his consciousness to his molecular make-up may be, there will have to be a continuing effort to understand learning in the cognitive, affective and motor domains as set out in Bloom's taxonomy.

There seems to be little doubt that many of the fruitful endeavours to examine man's ability to learn will be biological in nature and that the analogy to the computer will become ever more exact.

CHAPTER ELEVEN
SOME IMPLICATIONS OF CONTRACTING-OUT RESEARCH—1978

My purpose in this document is neither to defend nor to criticize the government's "contracting-out" policies, but rather to look to the future and to examine some of the problems that the government and the scientific establishment of a modern industrialized state must inevitably face. As Stafford Beer quotes from Warren S. McCullogh: *"Don't bite my finger—look where it is pointing"*.

The original paper was written as a discussion document for a management/staff committee, whose mandate was to discuss and anticipate areas of concern or possible conflict between the scientific staff and management of the National Research Council of Canada in Ottawa. It caused some very strong positions to be taken by the staff, some of whom felt that they had been betrayed from within. Letters were written to the President, to Dr. Hertzberg and to me. They did not heed my opening quotation "Don't bite my finger—look where it is pointing" and they failed to realize that the annotated references cited, were there to represent the view of "science" and by implication, the view of the NRC, as seen from the "outside"; these views being the ones that would most strongly influence the government's interactions with the NRC.

The current [late 1986] budget cuts imposed on the NRC by the current Conservative government have emphasized my original concerns and these notes are intended as explanatory comments. I make no distinctions regarding the political bias of the government in power.

The Global Problem

The body of scientific knowledge is never complete, but our knowledge of fundamental processes or "natural laws" is probably now sufficient to allow mankind to achieve solutions to many of the pressing global problems. The major deficiency is in the organization and utilization of that knowledge and in the political or social processes whereby that knowledge is applied.

In the long evolution of life on this planet, knowledge has only become "knowable" with the recent arrival of man, along with his self-awareness. In terms of the "Cosmic Year", Galileo came on the scene on December 31 at 11:59:59. Within this quantum jump in knowledge there are several definable inflection points. The ability to codify and systematically record knowledge led to science, i.e. knowledge about the natural world. The application of the inductive method of reasoning fostered by Francis Bacon opened the flood gates to the technologically derived industrial age. We are now entering what is called the post-industrial or quaternary age.

In this section I am trying to show that it is our political, economic and social systems that are deficient in solving human problems. Science in its most general sense has provided a massive fund of knowledge in a very short time, when viewed over the course of recorded history. The sophistication of that knowledge is increasing at an exponential rate, but we are falling behind in the application of that knowledge in ways that enhance the quality of life.

Insecurity in the face of possible annihilation due to nuclear war does not enhance the quality of life, in spite of the material comforts of modern "western" society.

One of the problems in dealing with the transition from the industrial (tertiary) age to the quaternary age is to confront the fact that we tend to think and act in a linear fashion. "Today is like yesterday and tomorrow will be like today". We are unaware, in our daily lives, of the exponential rate of change. One effect of the explosive development of technology is to compress the available time scale. The rate of change has itself changed. Another effect is that both goods and ideas can be propagated very rapidly on a global scale. On the other hand, the time taken for a new idea to become effective in the marketplace does not seem to be reduced to the same degree. In other words, there are many problems facing us that will have to be dealt with using current knowledge.

One of the major deficiencies in our current problem-solving activity is the organization and distribution of that knowledge. We have acquired technological knowledge and we have used that knowledge to control human activities—in other words, the human control of our interactions with each other is now of higher priority than the development of technology. We have "over produced" in a technological sense vs. the organizational processes.

There is need to examine the term 'scientific knowledge'. Western countries tend to regard it as knowledge obtained in the "laboratory", yet mankind has accumulated vast stores of knowledge in the form of "folk lore" which, over millennia, has been used to enhance the "quality of life" of many diverse societies (the role of the medicine man).

The explosive output of human activity has produced interactions and relationships, both of people and information, which have become so inter-woven that the resulting complexity is almost, if not totally, out of control.

The term "complexity" has been used to mean the related political, sociological and technological interactions in their totality that take place in any well-defined national entity. The management of complexity has enabled the multinationals to operate on a global scale—albeit for very specific economic and industrial purposes. Also, the processes necessary for the management of complexity (read "organization") is probably one of the most significant outcomes of the US and USSR space programs. The link-up of the American and Russian space vehicles testifies to the effectiveness of international management applied to a very complex task. To reiterate: The post-industrial age faces the challenge of the management of complexity in the very complex area of human relationships.

The Canadian Context

Concerns for global economic, social and industrial policies and problems indicate that these problems and related questions are in the process of formulation. Because of the time-scale compression mentioned earlier, governments are reacting with ever increasing concern to the impact of new and troublesome problems. In Canada, we are reacting also to both global and national technological and social problems.

The federal government reacts (by definition, a governor reacts after the fact or stimulus) to external pressures—demographic, social, political, etc. A government has no "divine" source of knowledge and so must seek its way to solutions of problems using the knowledge available to it. Also, no modern government operates in isolation from the rest of the world.

Given our natural resources and the increasing pressure of the explosive use of primary products, we are trying to develop an industrial base in Canada so that the industrial output derived from these products will be to our economic advantage. We are also slowly becoming aware that conservation of these resources is critically important.

The federal government, which has responsibility for all Canadians as well as those who are scientists, has long had an expressed concern regarding our industrial base (industry tends to regard this concern as a paper blizzard!). Since NRC is a component of the federal government, we should not be surprised that NRC has become part of the apparatus used by the government in its reaction to the global situation I have attempted to outline.

Canadian governments have never given the science portfolio any significant priority in their cabinet hierarchy. It is usually given to a junior and inexperienced minister and the incumbent is often shifted long before becoming familiar with the science community and its problems. For this reason (among others) the government's use of science in its policies is, at best, perfunctory, and at worst, downright destructive of the scientific community

The need to foster an inventive, intelligent use of scientific knowledge led to the formation of NRC in 1916. Reasonable men in government are also aware that the source of scientific knowledge comes from the dedicated searching of scientists, i.e. from "pure research".

What they don't understand is the process. We, within NRC, may understand the process but by the same token we may not know much about the application of that knowledge in the "real world". The development of a research environment in Canadian industry has long been a National Research Council priority.

Perhaps we should also recognize that in the post-industrial age—the age of the management of complexity—we have an even greater priority and that is how to organize and disseminate that knowledge for utilization in an industrial, economic and sociological context.

One of the federal government's approaches to these problems on a national scale (but remember; the pressures are international) has been to recognize that:

".... science and technology are vital to the physical and social structure of Canada and, if wisely fostered and used, can provide desirable future options for Canadians. Since the government has many science and technology requirements to meet in support of departmental missions, its policy of contracting-out these requirements reflects its belief that it is in the national interest to encourage the fullest possible participation in Canadian industry in meeting these needs, to stimulate industrial innovation and thus provide additional benefits to the economy. The government has therefore provided, subject to overall financial constraints, that its mission-oriented, science and technology requirements in the natural sciences and the human science fields of urban, regional and transportation studies be contracted-out to the private sector, and especially to Canadian industry. This is done to obtain a more even balance in the coming years between scientific activities performed by industry and by government in support of department missions, except in specific limited circumstances as prescribed below. The significance of this policy to departments will depend on the degree to which departmental practices already conform to it and on the extent to which departmental S&T requirements are excerpted under it." From: "Policy and Guidelines on Contracting-out the Government's Requirements in Science and Technology", Administrative Policy Branch, Treasury Board, April 1977.

This approach, while laudable as presented on paper, has not worked to the advantage of industrial Research and Development (R&D). The percentage of industrial R&D has not changed appreciably over many years. NRC was making pleas for more R&D in the industrial sector over thirty years ago.

What will produce a change is an investment in R&D that is commensurate with that of other industrialized nations. The present government has paid the usual lip-service to R&D, but has, in fact, reduced its support, while building unnecessary prisons (the role of education, poverty and despair is another story).

The spin-off from a massive investment in R&D is what produced "Silicon Valley" and "Route 128" plus the presence of nearby universities of recognized excellence. We in Canada do not have those resources; nor will we ever have them if we follow present government policies of putting science on the back burner (or even occasionally turning it off altogether).

The most extensive (though flawed in the opinion of many scientists) study of science in Canada was the Lamontange Science Policy study done in the 1960's. How many current members of Parliament know of its existence or recommendations?

The criteria listed on page five of the Guidelines do not deal with the provisions for maintaining a first-class scientific establishment committed to excellence. In Appendix 1 under Definitions, item five (page 13) states in part: "Free basic research ... is not generally intended to be included in mission-oriented science and technology. Other more appropriate mechanisms already exist to provide for performance in the private sector (e.g. programs for grants, fellowships, etc.)

NRC seems to be reacting on two fronts to the government's decision to "contract-out". The response of management has been outlined in the document *"NRC—A Redefined Role"* extracted from the 1974-75 Report of the President and recently distributed to the staff. While not "contracting-out" as defined by the government, NRC's involvement in the government's IRAP (Industrial Research Assistance Program) and NRC's operation of PILP (Program of

Industry/Laboratory Projects) show that NRC has responded to the spirit and intent of the government's mission-oriented science and technology policy.

Given the premise of this paper, the government's involvement of NRC in its perception of Canada's economic and industrial problems is inevitable.

One result will be that governments will thrash around looking for ways to deal with these problems, but without a strong commitment to research, they will at worst fail miserably, or at best, solve some and create more.

And now [November 1986], the current Conservative government has appointed a commission to tell NRC, for the first time in its history, how to manage its administrative and fiscal affairs. The result will be catastrophic in its effect upon productive and original research. (See my Postscript below.)

The government seems to be oblivious to the fact that, after spending a lifetime in being educated in science and pursuing a single minded goal in specialized research, a scientist cannot just "be transferred" to another discipline and be expected to compete in that new discipline with the best brains in the world.

The staff is responding to the contracting-out policy and also to NRC's response to that policy. But since the staff have not been directly confronted with the implementation of the policy, their knowledge of "what" and "why" may not be particularly well detailed.

It is probably a valid generalization to say that the professional staff regard their work as significant and useful. More particularly, most of the staff is convinced that there is always a need for new knowledge and they can cite many instances of serendipity in the process of scientific inquiry with the subsequent use of that newly acquired knowledge being used to solve old problems.

Again, the premise of this paper is that the complexity of global problems has increased and the time scale in which solutions can be effected has been compressed; also the nature of the problems has changed, from one of the need of technical solutions to the management of complexity.

Thus it would seem that all governments of all nations have no choice but to utilize all the resources they can intelligently muster in order that those nations survive. That is not to say that new knowledge is not needed, even desperately needed, but it would seem that the government no longer regards the privilege of the scientist to pursue his "nuggets of truth" as sacred. That privilege is for the very few—and the selection process whereby that privilege is accorded must be even more rigorous. [This comment was misconstrued and resulted in many angry letters to NRC's President.]

An increasing proportion of NRC's activities therefore should be devoted, not only to solving problems of national "importance", where solutions mean technological fixes, but rather in organizing the flow of scientific information to, and for, the government with overwhelming logic so that the national machinery—industry, trade and commerce—can be used to make the inevitable transition to a steady state existence in equilibrium with both resource and energy sources.

What I am saying is that a new facet of the mandate of institutions such as NRC and universities, is to examine, not only the role of fundamental research, but also to establish the infrastructure which can examine complexity to both extract and explain the essential information in such a way that effective corrective social and political decisions can be taken.

The current democratic process of election of members of our society to provincial and national governments must be changed to invoke selection criteria, which ensure that those chosen provide a government whose reactions to problems are based on current knowledge and not just on legal or economic intuitions or partisan political bias.

In Summary

1. Canada is inevitably affected by the global explosion of technology.

2. The time scale for the solution of problems has been compressed.

3. Many problems will require the intensive application of existing knowledge—the final solution can't wait.

4. Canada is part of the global matrix—the federal government responds as a part of that matrix.

5. NRC as a department of the government must be prepared to be asked to play a part in the solution of the government's perceived problems.

6. The government *seems to perceive* that the luxury of "dabbling" in pure research may well have passed; therefore pure research is only for the exceptionally gifted.

7. The "assault on complexity" may well have to be one of the major NRC tasks.

8. Pressures other than "contracting-out" can be expected and the staff must give management support in both anticipating and responding to these pressures—there will be an increasing demand to devote effort to organizing existing knowledge or maximum utilization in solving national problems.

9. Management must, by the same token, maintain effective communication channels with the staff.

10. The organization of NRC must be viewed as fluid over the next several decades to meet these challenges.

REFERENCES

The references cited make my point and I have no reason to believe that the situation has changed in any substantial way. These references have been culled from my own files and they are cited to indicate both the global and the particular challenges that face the National Research Council. They have been omitted since they were used to emphasize the pressures on NRC. I include Reference #7 for its unique and now, timely import:

> 7. Markov, M. A. *Science and the Responsibility of Scientists* , Impact of Science on Society, Vol. 25, No. 1, 1975, p. 19.

> Opens with a quote from Pliny the Elder written 1,900 years ago:

> We must give an account of the metal known as iron, the most useful and the most fatal in the hand of mankind. .... But it is with iron also that wars, murders and robberies are effected .... This last I regard as the most criminal artifice that has been devised by the human mind; for, as if to bring death upon man with still greater rapidity, we have given wings to iron and taught it to fly. Let us therefore acquit nature of a charge that belongs to man himself.

POSTSCRIPT

My concerns were prescient—the NRC we knew in those golden post-war years is no more—we could visit and talk with anyone at NRC. I retired in December, 1979 and moved to Calgary in 1988.

In 1992, Cumberland Township invited me to attend a ceremony recognizing my contribution in developing the Township Heritage Museum. I took advantage of that visit to see one of my friends at NRC. I was astounded to find myself confronted by a solid glass wall as soon as I entered the building where I used to work. Research work conducted on behalf of industry demand complete secrecy!

I was given a visitor badge, my friend took me through the locked glass door to his own laboratory; he then returned me directly to the outside world as we said "Goodbye". Prime Minister Harper has recently appointed a non-scientist as President of NRC—the destruction is complete.

The Government's position is that pure research is a function of the university community, except that much of that research is also now funded by industry.

CHAPTER TWELVE
SCIENCE, TECHNOLOGY
AND SOCIETY—1986

This chapter is a combination of three essays, each of which dealt with some aspect of the topic titles. All three derive from those initial concerns presented in my talk delivered while I was at the University of Western Ontario in 1946: Science, Education and Catastrophe.

It is a moot point whether Religion, Science, or War, has had the greatest influence on the development of societies. Certainly, whole societies have evolved around a particular religious belief; those societies have been modified by acts of war among themselves; acts of war have also changed societies by the application of scientific knowledge to the technology of war, and subsequently the evolving technology of that society.

While in my first year at the University of Western Ontario in 1946, during a course on Public Speaking, a class assignment was to prepare for a talk, in which we were to convince our audience of the validity of our point of view about a topic of our choice. My topic was *Science, Education or Catastrophe*. My thesis was that, unless the public at large understood the significance and processes of scientific enquiry, the unfettered uses of technologies derived from scientific knowledge could have catastrophic consequences.

John Ziman, in his book, *The Force of Knowledge—The Scientific Dimension of Society* (1976), says in his opening chapter:

> "This book is not a sermon on these moral issues. Each of us, as a responsible citizen of the world, must find his own answers to such controversial questions. But, to think constructively about these matters, it is necessary to know a little bit about the nature of science as a human activity. It is not sufficient to understand the discoveries that scientists have made around the world; we must also learn to see scientific research as an integral part of the modern way of life. Rational debate on the political and moral issues concerning science and its place in society should be staged against some background of facts and agreed principles. The aim of this book is to sketch out a background for such debates."

His concluding sentence in his book is:

> "This is the reason why a certain amount of exercise of mind and spirit on these difficult and multifold issues should be part of the education of every scientist—perhaps of every thinking person."

In the discussion that follows, I am not writing a scientific thesis about the influence of science and technology on society. I intend to summarize some of what is known, and then present my own interpretation of the consequences. Some deleterious consequences of technological advances:

Summary

1. Huge mechanized fishing trawlers have virtually extinguished some ocean fish stocks.

2. Greenhouse gases have brought on global warming with consequences we are just beginning to comprehend.

3. International air travel has hastened the spread of infectious diseases around the world.

4. The development of the transistor has spawned modern computer technologies that have changed societies; even those that are considered underdeveloped.

My Interpretation

1. Those who have promoted the development of mechanized fishing trawlers were almost certainly concerned only with maximizing their profits and supplying the food market with fish at a lower cost than their competitors. If they had any concerns about the depletion of fish stocks, they proceeded anyway.

2. Warnings about global warming are not recent; tree-ring records from Mongolia indicate that the 20th Century is the warmest in the past 1000 years. It has only been widely recognized in the past few decades that global warming is in fact occurring at an increasing rate. The latent energy increase in the atmosphere due to a temperature increase of one degree is enormous, as is the ability of the warmer air to retain water. Severe wind storms, and abnormally heavy rain, are consequences of higher global temperatures. Receding ice in the arctic regions is another indication that temperatures are rising. Melting of the arctic ice will result in flooding of coastal cities around the world. The Kyoto Protocol was designed for a world-wide reduction of greenhouse gases in a controlled, progressive manner; it has been rejected by the largest single user of technologies producing such gasses, namely, the United States of America.

3. Currently, a deadly avian flu virus in China is causing great concern that air travel will spread the virus around the world and cause a world-wide pandemic. AIDS was first detected in North America when an air steward was diagnosed in New York. The SARS virus was brought to Canada by an air passenger. With bigger and faster airplanes the possibility of viral infections spreading will be increased. AirBus in France has just announced an 800-passenger plane that has passed it initial flight test. The cost per passenger-mile will decrease, but what are the global consequences?

4. When John Bardeen and Walter Brattain were probing the surface state of crystals in 1947 and found that one of two closely spaced sharp probes could influence the current flowing in the other, they were unaware of the technological consequences. They regarded the effect they observed as that of a voltage on one probe affecting the resistance of the other, hence the name transfer resistor—transistor. It took another twenty-five years for the transistor to evolve into modern LSI (Large Scale Integration) technology. Once discrete components could be formed on one crystal wafer and connected internally to function as a complete integrated circuit, with external connections for the power and control functions within a two inch-square encapsulated case, the floodgates were opened.

This occurred in the early 1970s, now, in 2005, the CPU (Central Processing Unit) of a desktop computer can contain many millions of transistors and solid state components, with a computational speed that exceeds a billion cycles per second. This necessarily brief, and inadequate, description is included only for the purpose of introducing the concept of technologies that are at the heart of almost all modern commerce and communications. If all CPU functions were to be shut down, almost all economic activity in the developed world would collapse.

One of the causes of concern about the proliferation of nuclear weapons is that a nuclear bomb explosion in outer space would emit a disastrous electro-magnetic pulse. That pulse could instantaneously disable all electronic devices by inducing overload currents in conducting materials in electronic and other devices within the portion the of earth that is within line of sight. I expect that there is no public mention of an electro-magnetic pulse, either because some way has been found to counteract the pulse, which I doubt, or because such public knowledge would cause consternation. Such information is available on the Internet.

These four examples are sufficient to indicate the unintended, or unknown, consequences of the technological application of scientific knowledge to problems that have technological solutions. That the activity may have deleterious consequences, at a later time, is either unknown, outside the scientific world, and is not considered—if it is known—it may be ignored, or that knowledge of potential consequences does not yet exist: hence; *Science, Education and Catastrophe.*

People who are in a position of public trust, that is politicians, have a responsibility to be informed, or to listen to deputies, whose job is to be informed. Because they wield so much power, CEOs of large international corporations have a similar moral responsibility. Unfortunately, their undeclared responsibility is not to the public good, but to the survival and growth of the corporation.

Each one of the four factors listed above have other aspects that affect society. When they become commercially viable, technologies derived from scientific knowledge become an economic component of society. That is not to say that the bulk of the people, who are affected by their introduction into the fabric of society, demand their use, but that there is an economic benefit to corporations from their use.

Thurman W. Arnold in his book *The Folklore of Capitalism*, quotes from Joseph Dorfman's book *Thorstein Velben and His America* as follows:

1. "God has made man a creature of desires" and has established the material universe "with qualities and powers … for the gratification of those desires." Desire is the stimulus to production and invention.

2. To satisfy desires, to obtain pleasures, man must by "irksome" labour, force nature to yield her hidden resources.

3. The exertion of labour establishes a right of PROPERTY in the fruits of labour, and the "idea of exclusive possession is a necessary consequence." Originally the object belongs to the producer "by an intuitive conception of right, and the act of appropriation is as instinctive as the act of breathing." The right of property may be conceived as "a law of natural justice," as Bowen of Harvard put it, because "the producer would not put forth his force and ingenuity if others deprived him of their fruits." Thus is established:

4. The right of EXCHANGE.

   "Here is the beginning of the religion of the essential dignity of an individual's accumulating wealth by trading which later became the mystical philosophy that put the corporate organization ahead of the governmental organization in prestige and power, by identifying it with the individual. Our fathers breathed this atmosphere in every day of their schooling."

Arnold goes on to quote from Dorfman:

"Since socialism is the utter negation of the right of private property, man is no more adapted to it than the barn owl is to live in the water. Philanthropy or any other aid to the poor is a violation of the same laws of God and property. All attempts to relieve the natural penalties of indolence and improvidence bring about unexpected and severe evil. The doctrine that the government should provide for the unemployed is the most subversive of all social order. Even the claim of Ruskin that all labours of like amounts should receive the same reward, means the suppression of commercial law, which is God's method. If labour and capital are free, as they are in the order of nature undisturbed under the law of competition, then the flow of each °K toward an equilibrium, is as natural as that of waters of the ocean under gravitation. In reality the labourer has no complaint against the competitive system. As Perry put it, employer and employee come together of necessity into a relation of mutual dependence, which God has ordained. And which, though man may temporarily disturb it, he can never overthrow."

As I interpret Arnold's quote from Joseph Dorfman and Thorstein Velben, the mystical philosophy, or religion, that guides the acquisition of private property, is based upon the dependence of the employee on the munificence of his employer, that is the owner of the organization, as God has ordained. It does not mean that the employee's social well-being is of any concern to his employer, except that his health may be a necessary requirement in his employer's work force. The laws of God are also the laws that govern the behaviour of the employer in his capitalist environment, which is sacrosanct to him above all other laws of state or country. The true faith from which the laws of God are derived is Capitalism.

Also, private property is an euphemism for an organization known as the Company, or the Corporation which, as Arnold has pointed out, is legally regarded as an individual in commercial law. The Oxford Universal Dictionary defines a corporation as:

> A body corporate legally authorized to act as a single individual; an artificial person created by royal charter, prescription, or legislative act, and having the capacity of perpetual succession (1611). One definition of capital is defined as: Accumulated wealth employed reproductively (1630).

An incorporated organization can exist in perpetuity. The ownership, management and function can change with time, but the corporate individual continues to exist until it is legally dissolved. Corporations have now become "globalized"; they function internationally across country boundaries.

Now, back to my concerns.

The purpose of an owner(s) of a corporation is to ensure its viability, growth and profitability at all times; that role is, I believe, assigned to the Chief Executive Officer or CEO. I have no direct knowledge of the actual operation and management of any company or corporation.

Alfred P. Sloan, Jr., in his book *My Years with General Motors* (1963) has entitled Chapter 19, as: Nonautomotive: Diesel Electric Locomotives, Appliances, Aviation. In his introductory remarks he said:

> "It would be nice to be able to trace a coherent pattern in General Motors ventures outside the automobile business, but chance and other factors that entered the picture make it difficult to do so. We had, of course, some natural interest in diversification which might afford us a hedge against any decline in automobile sales. But we never had a master plan for nonautomotive ventures; we

got into them for different reasons, and we were
very lucky at some crucial points. We got into the
diesel field, for example, because of Mr. Kettering's
special interest in diesel engines, dating back as
early as 1913 …"

Perhaps Mr. Sloan was thinking about the way GM increased the
sales if its buses and cars in Los Angeles and other major cities in the
United States.

Donald C. Carr, in his book *Energy and the Earth Machine* (1976),
had this to say on page 210:

> "No court in the land had the starch or guts to
> suggest even the possibility of GM's huge plot—so
> huge in fact that one can see a certain legal gen-
> eralization emerging: If your conspiratorial trust is
> big enough, if it involves practically every citizen's
> pocket, then the courts will do little or nothing,
> because they come to accept it all as the directed
> circumstances of God in a God fearing nation."

What Carr is writing about is the replacement of the electric street
car system in Los Angeles with a massive road system for cars and
buses. According to Carr, General Motors planned to replace the
existing electric streetcar systems in major cities, and then have them
replaced by roadways for cars and buses. By 1938, GM had designed
a new bus body and a new diesel engine to power them.

Their next step in the take-over program, was to incorporate
National City Lines, with subsidiaries which operated in specific
cities. Pacific City Lines operated in Los Angeles and San Francisco.
By 1940 Pacific City Lines were buying up the Pacific Electrical
Company in Los Angeles and the Key System in San Francisco. By
1951, the last big red car ran through Watts to Long Beach; "along
with its hundred brothers it would be stripped and burned. The

unparalleled 1,164 miles of broad-gauge track was torn up and sold as scrap steel."

Since GM had the monopoly on bus production they could decide just how many buses would be made available to replace the street cars. The street cars were replaced by fewer buses and many more cars. Los Angeles became the smog capital of the world. In all, fifty-six major cities in the United States had their street car systems replaced by National City Lines and their subsidiaries, with buses, and of course, with many more cars. Those cars drivers demanded more and better roads for their cars, all at the taxpayer's expense.

This is a blatant example of a case where both the public and governments were exploited for the economic benefit of a large corporation that declared in its public statements it was providing what the public wanted. Yet all the while it was creating public desires for bigger cars with annual model changes, an increased number of new models, all with bigger and more powerful engines.

None of which was necessary if the fundamental function of the car was transportation from here to there. In that case, what is necessary is increased safety of the body design, improved longevity and efficiency in the use of fuel. It would be many years before those factors were seriously considered in automotive design. Even now in the year 2005, automobile manufacturers are proudly advertising bigger, better, newer cars with advanced features such as audio systems with 15 loudspeakers; the biggest sellers are the SUVs (Sports Utility Vehicle) having an increased fuel consumption, while the press headlines are saying that we will soon be running out of oil.

Not only are we running out of oil, but the uses for that energy are changing. Oil is fueling our machinery; oil and natural gas are the sources of the long polymers that are used in the manufacture of many plastics. They are also discarded as waste in our garbage bags. So much of our consumables are made by automated machines— once they are running all that is needed is to feed the machine with its appropriate diet of material and ship the packaged output to Wal-Mart. The machine does not care that it has displaced the workers who are destined to purchase its output.

In 1986, I had been listening to a discussion about unemployment between Peter Gzowski and Ken Dryden on CBC *Morningside*. Ken was saying that unemployment would be a permanent problem from now on—we would never again have full employment. I sat down at my KayPro II and wrote the following essay.

The missing element in Ken's analysis is the role of complexity in the modern industrial world. The inter-connectedness, inter-dependencies and the speed of communications within that world have fundamentally changed its mode of functioning. Regarded as an entity, that complexity has evolved into a "thing" in its own right. It is now so pervasive that we, that is people, both individually and collectively, are no longer in control. "Complexity" is in control, Eisenhower warned of the "Military-Industrial Complex"; true, his warning was in a different context, but his intuition was correct.

Power had migrated to a complex system that was not deliberately created, but nevertheless had come into being through the growth of ever larger integrated machine-like systems, supported by ever larger public funding. Now the American public funding of their system is approaching a trillion dollars—an amount beyond the control or conception of any individual. True, the figures may be comprehended, but the effects cannot.

Let me coin a word for this complex system to make the writing easier; more thought is needed to devise a proper term, but for now I suggest "Cmachine" as a contraction of "complex machine".

This Cmachine has a life of its own and we, that is people, have become totally dependent upon it to sustain us. We no longer depend upon our local environment for our life support. The other aspect of the Cmachine is that it is increasingly capable of supplying its own needs through automation; that is it does not depend solely on human labour to sustain it. CEOs and Boards of Directors are now committed to ensuring its survival by reducing labour costs and increasing productivity through automation. People are becoming redundant, or in other words, unemployed. If this analysis has any validity, unemployment is a permanent fact of modern industrial society, and it will increase over the long term. We will never again see full employment.

Now the question is what to do about the Cmachine?

First, we have to realize that we cannot return to simpler times, except through catastrophic disaster. What, then, can we do? Perhaps we can consider how we distribute the output of the Cmachine. We still have some control in this area, at least governments do. Currently, we depend upon an economic system for managing the distribution of its produce. If we work, we share in the output; if we don't work then we depend upon government support and charity. The Cmachine doesn't care since it has no compassion. Perhaps we need to recognize that it is up to us to make this Cmachine serve our needs through making it into a basic life-support system available to all people by government intervention.

This concept will require a book to examine it in any detail.

Perhaps we need to ask these questions; What is the purpose of the conversion of the Cmachine's output into entries in a bank account ledger? Is it because nature decreed it that way? Is it because the machine demands it? Is it because it enables governments to keep their GNP accounts in order? Or is it, perhaps, that it enables individuals or compact groups of people to exercise considerable power and control over major portions of the economy and other people; and, do they assume this control is theirs by divine right?

Let us ask another question: Is it a natural right to expect that the environment in which we live, provides us with the essentials to sustain life? The fox catches her prey and devours it in order to live and perpetuate her species; all natural life appears to be sustained by a complex food chain. And now we seem to have become the penultimate link in that chain, with the Cmachine as the ultimate entity devouring all. Perhaps this view is too cynical, but if it focuses our attention on a different conception of our relationship with our industrial and agricultural technologies then it may be useful.

Ken talked about the differences in our generalized concept of unemployment and that of the Swedes. All levels of Swedish society are implicitly committed to the concept of full employment. It is doubtful that the Swedish economy is as devoted to the industrial machine as is that of North America; also, it is much smaller.

Perhaps we need to re-examine the capitalistic-free-enterprise foundation of our economy. Certainly, current professional economic theorists will try to analyze, interpret and modify this system from their well established bases. That however, does not say that a radically different approach may not lead to an understanding that may produce modifications which could provide life-support systems more in keeping with the natural evolutionary support system that was in effect before we modified it as we developed the Cmachine.

Any modification should provide for at least two fundamental human requirements; the need for food and shelter for all members of society and the need for the exercise of human ingenuity and creativity. It is a tenet of human behaviour that if all wealth were evenly re-distributed, it would very quickly become concentrated again in a relatively few hands, because of the inevitability of human greed and the need to exercise power. Because of this tenet it would require a covenant between all people through government decree that the Cmachine output would have to provide a basic life-support base for everyone.

Currently, the idea of a negative income tax recognizes that such a concept is being considered. Universal health care systems, old age pensions, unemployment insurance and other support systems are tendencies in this direction. The concept of an industrial eco-system for general life support is not part of our general thinking; in fact I am not aware of it ever being discussed. The "free-enterprise" idea of ownership of material possessions implies that the output of the Cmachine is generally regarded as being owned by the "company". Perhaps the recognition that none of its output would exist without the input of people should imply that a share of that output should be made universally available.

Wages and taxes may be regarded as one way to do this; wages are often inadequate and taxes are not often adjusted so that the level of life support becomes a right of human existence as is the air we breathe; and taxes are subject to cynical manipulation for individual or company benefit.

If one extrapolates this concept into the future then some very perplexing questions arise. We must eventually arrive at a population density and a productive technology that is in balance with the ability of the planet to support that population. How will this balance be maintained? Will it be through a "survival of the fittest" regime or will some humanitarian and ecologically responsible solution be found?

Whatever the modification, there remains the other question; what are the unemployed to do? Creative use of one's talents seems to be one of the most satisfying of human endeavours. Perhaps educational policies will have to be modified to place more emphasis on this aspect of preparation for a lifestyle in the new age of living with the Cmachine.

This also requires another book!

We have had several decades to consider the question of unemployment; from inhumane treatment during the "dirty thirties" to modern unemployment insurance and welfare support. What we have not done is to fully understand the inevitability of the impact of the complex industrial, economic, and increasingly computer-automated society, on the lives of all levels of that society, nor have we addressed the need to respect the dignity of each individual person.

Unrest, riots, soaring divorce rates, food banks, terrorism, suicide among the youth, pollution of the environment and a host of other deleterious social effects may well be a part of the penalty for introducing a massive and abrupt change into what was a slow evolutionary process.

Do we still have the time or the wisdom to examine what we as a society are doing collectively, not only to that society, but also to the environment in which that society functions? The explosive use of cheap, portable energy has transformed the world everywhere. But those energy sources are very limited in the long-term sense. Do we have an ultimate responsibility as trustees? Are we holding the fate of the future in our hands today? We are here today only because of what has gone before; the perceptions and endeavours of our ancestors.

On a time scale that stretches from several thousand years BC to several thousand years into the future, our energy use is as a small candle glowing briefly in the darkness during the period we call the present. Given a "cosmic year" time scale equivalent to the four-billion-year age of the universe, the period of human history that we have been discussing occurred during only the last few seconds of the last day of that cosmic year.

Lord, forgive us, for we know not what we do."

CHAPTER THIRTEEN
ON WATCHING "EARTH" TVO,
November 24, 1987

This was followed by the First Ministers Economic Conference on November 25. My emotional involvement in these issues was so intense that I had difficulty dealing with it. For many years—at least 40—I have been thinking and to some extent writing about these issues. By this time [1980s] I was well aware of what mankind was doing to his only home—Planet Earth.

Thames Television has prepared a program called, simply, EARTH. It is dedicated to Harrison Brown who wrote "*The Challenge of Man's Future*" in which he, for the first time, described the inevitable disastrous consequences of man's population growth and the need to control the impact of his technological, industrial and economic impact on the only home we have; planet EARTH. I have his book—it is one that I cherish.

I had not seen Harrison Brown before. The film was obviously made toward the end of his life (he died in 1986). The image he projected was of a care-worn, kindly person who spoke quietly, but from deep conviction. Among the others interviewed were; Rachel Carson (*Silent Spring*), Paul Erlich (*The Population Bomb*) and Donnalea Meadows (*Limits to Growth*). Each one projected a strong image of openness, kindness and concern. Why is it that politicians seem to respond only to characteristics that represent power and control? There are some notable exceptions, such as Tommy Douglas and Stanley Knowles in Canada.

My emotional involvement in this issue is so intense that I have difficulty dealing with it. For many years—at least 40—I have been thinking and to some extent writing about these issues. I do not pretend that I have written anything very profound, but the perception of the inevitable consequences has been so very clear to me that I felt that, even though my formal role as a research scientist with the National Research Council demanded almost all of my time, I had to give what I could of my "free" time to the consideration of ways to make mankind's transition through this dramatic and traumatic time at least a little more probable.

To this end, I have spent much of my effort in the area of science education, because it was in this area that I had some credibility and where I could perhaps enhance the incremental increase in relevant knowledge of the young people. They are the ones who will have to produce major changes in mankind's habits, attitudes and global interactions. Remember "Two Generations of Teachers Away from Disaster"?

The scenes on TV of the recent stock market crash and the continuing destruction of the world's forests, particularly the Amazon rain forest, indicate that man's greed is still in control. Governments, while relatively effective in maintaining economic and productive activities in some sort of coherence, do not seem to be particularly aware of the planet EARTH. Sure, they give a nod of the head to such concerns, but the exercise of power and maintaining man's, rather than Nature's, mode of operating is by far more important to them.

Experiments have been conducted using rats to examine the role of "pleasure centres" in the brain; electrodes were inserted into these centres in the rat's brain and the electrical stimulus could be applied by the rat itself when it pressed a switch. The researchers were astonished to observe that the rats found the experience so pleasurable (addictive?) that they would prefer the sensation to eating and they would have died rather that give up their pleasures.

Sometimes I think mankind is like the rats—he is so addicted to his own exploitive pleasures in the search for power that he is indifferent to the knowledge that he now has the potential to "kill" himself in the pursuit of those pleasures. This is why I feel that the exploration of a different system of governance is imperative. We must get our addiction under control. Our present "democracy" seems to ensure that we pursue those pleasures with ever more intensity. Movements such as the Green Party, Greenpeace and others (there are many around the world) do have an impact, but a bottle of beer and the sports channel on TV seem to occupy a very large portion of the male population, while the "soaps" and game shows on TV seem to attract the females.

In the "underdeveloped" parts of the world none of the above is relevant; just living is an all-consuming task. Yet the demands on the life support systems are as crucial there as anywhere else. The wisdom of the ancient "medicine man" and the tribe elders long ago seemed to support an ecologically sound relationship with the natural world around them. The mysticism and rituals were a method of encoding that knowledge and ensuring that it was passed on to succeeding generations. We seem to ignore all that for selfish self-gratification and the exploitation of both people and resources. All this has been said many times in many ways; why is it not understood and acted upon?

November 25, 1987

I watched the First Ministers Economic conference today. The opening speeches were broadcast in their entirety. As expected there were the predicted divisions of opinion on the Free Trade deal. But in all cases, with the possible exception of Premier Pawley from

Manitoba (his rhetoric is so convoluted that it is hard to follow him!) they all want the benefits of increased trade without the negative consequences; some of them do not see any negative effects at all (Vander Zalm, Getty, Devine and Bourassa).

In my comments on the film EARTH, I mentioned the politicians involvement with power rather than compassion (not explicitly in these words, but implied). This conference is a distressing example of that obsession. The discussions were concerned with the benefits of increased trade, and the resulting increase in job opportunities that would flow from an increase in the "money" available to run the economy and provide the services needed in a complex modern society. The expressed (or implicit) need for growth was regarded as the essential key to progress and human happiness.

What was not discussed, implied or recognized was the inevitable consequences of living on a finite planet, with finite space and resources. I'm all right Jack—too bad about you. To me. this conference represented the distressing scene of people given the responsibility for the destiny of our nation, playing their games in public and not being even aware of the nature of the game they are playing.

My comments about sports and the bottle of beer were a bit too close to the mark to be comfortable. Several Premiers mentioned football in developing their analogies, beer and wines also played a part in the rhetoric, not as trade products, but as symbols in the discussion. All the participants were men. The rat's pleasure centres do have their counterpart in our society. Perhaps I am too sensitive and cynical, but I do see enormous problems if we do not back away from the American dominated headlong rush into the technological future—the leash length is finite and we will be brought up short with a terrible jerk some day.

CHAPTER FOURTEEN
THE EXISTENCE OF GOD—1988

While it was kept in the background of my daily life, religion has been a life-long concern. For this chapter I selected quotations from several well-known authors. I wanted to show how firmly religious concepts are held by intelligent people.

Probably the most consistent debate in the entire history of mankind has been about the existence of a Divine Deity or God, and if He/She does exist, how that existence is manifested. Those who take part in the debate always seem to accept the basic premise that God exists. No one argues that the concept of God does not exist; that belief is manifest around us, no matter what the terminology or the manner in which it is expressed. Even the most ardent atheist will accept the statement that most people believe implicitly in a Divine Being, or if you will, in God.

The following are expressions of man's relationship with his beliefs, as they refer to the concept of God. I present them to illustrate some powerful arguments in defense of: (a) religious belief in God exists outside and beyond science; (b) that religion and science can co-exist; (c) that one can believe that the Universe as a whole explains God, which in turn is not proof; (d) that God is busy elsewhere in the universe; or (e) that such beliefs are the product of mythology. I have chosen to use some quotations from an earlier work because they indicate that little has changed in the debate. Current atheistic arguments will enter the discussion in the next Chapter.

Lecomte du Noüy in his book *Human Destiny*, uses his scientific background to examine the religious roots of man's perceptions of himself, his place in society and his relationship with God. This is an eloquent statement of the situation 60 years ago—much has been discovered and learned since.

> Today [1947] Dr. du Noüy is known and respected by scientists of every land. In 1944, this respect was signalized by the University of Lausanne, Switzerland, when he was awarded the Arnold Reymond Prize, for his three books *Le Temps et la Vie*, *L'Homme devant la Science*, and *L'Avenir de L'Esprit*, as the most important contribution to scientific philosophy in the past ten years." (From Wikipedia)

In the Preface, he says, in part:

> "The problems of today have become so complex that a superficial smattering of knowledge is inadequate to enable the cultivated layman to grasp them all, much less to discuss them. This fact has been occasionally exploited in order to twist truth and to mislead the public. The time has come for all men of good will and of good faith to become conscious of the part they can and must play in life if our present Christian civilization is to endure. Everyone shares a responsibility in the future. But this responsibility can materialize into a constructive effort only if people realize the full meaning of their lives, the significance of their endeavors and of their struggles, and if they keep their faith in the high destiny of Man.

As the purpose of this book is to substantiate this faith by giving it a scientific basis, the writer hopes that the effort imposed on the reader will be rewarded by a clearer outlook on the most important problems of all times."

And on page 134, he summarizes his convictions regarding man's relationship with God:

"Once more we repeat that there is not a single fact or a single hypothesis, today, which gives an explanation of the birth of life or of natural evolution. As far as the origin of life is concerned, we have briefly studied the problem in the first part of this book. Willy-nilly we are, therefore, obliged either to admit the idea of a transcendent intervention, which the scientist may as well call God as anti-chance, or to simply recognize that we know nothing of these questions outside of a small number of mechanisms. This is not an act of faith, but an undisputed scientific statement. It is not we, but the convinced materialist who shows a powerful, even though negative, faith, when he obstinately continues to believe, without any proof, that the beginning of life, evolution, man's brain, and the birth of moral ideas will some day be scientifically accounted for. He forgets that this would necessitate the complete transformation of modern science, and that, consequently, his conviction is based on purely sentimental reasons.

Moreover, belief in God, today as in the time of St. Paul and St. James, consists in very little. A beautiful definition was given by a great Christian writer, Miguel de Unamuno: 'To believe in God is to desire His existence, and what is more, to act as though He existed.'

Many men who are intelligent and of good faith imagine they cannot believe in God because they are unable to conceive Him. An honest man, endowed with scientific curiosity, should not need to visualize God, any more than a physicist needs to visualize the electron. Any attempt at representation is necessarily crude and false, in both cases. The electron is materially inconceivable and yet, it is more perfectly known through its effects than a simple piece of wood. If we could really conceive God we could no longer believe in Him because our representation, being human, would inspire us with doubts. Of course this only applies to the man who is capable of criticizing his own intellectual mechanisms and of admitting the reality and value of intuition as well as of the irrational aspirations which, at an early stage of his development, spontaneously sprang up in the human being. These irrational aspirations are real. Man derives happiness from them and it has been wisely said that nothing which makes us happy is unreal. They are the source of our greatest virtues, of all our moral ideas, of our esthetic sense, of our thirst for ideals. Their cause must, therefore, be real also, even if it is inconceivable.

It is not the image we create of God which proves God. It is the effort we make to create this image."

Lest we misunderstand his position, he makes it very clear in the last few paragraphs of his book:

"It is only by direct action on youth that a better society can be successfully moulded. All pseudo-mysticisms social, philosophical or political must be replaced by the Christian mysticism, the only one based on liberty and the respect for human dignity. God keep us from judging. Humanity has not reached the age of reason and its efforts are still on the scale of the tribe.

Let every man remember that the destiny of mankind is incomparable and that it depends greatly on his will to collaborate in the transcendent task. Let him remember that the Law is, and always has been, to struggle and that the fight has lost nothing of its violence by being transposed from the material onto the spiritual plane; let him remember that his own dignity, his nobility as a human being, must emerge from his efforts to liberate himself from his bondage and to obey his deepest aspirations. And let him above all never forget that the divine spark is in him, in him alone, and that he is free to disregard it, to kill it, or to come closer to God by showing his eagerness to work with Him, and for Him."

Andrew D. Wright in his book *A History of The Warfare of Science and Theology in Christendom* tries to reconcile science and religion. White did not include any of du Noüy's insights; there is no mention of him in his Index. He says in his Introduction:

"My belief is that in the field left to them, their proper field, the clergy will more and more, as they cease to struggle against scientific methods and conclusions, do work even nobler and more beautiful than anything they have heretofore done. And this is saying much. My conviction is that Science, though it has evidently conquered Dogmatic Theology based on biblical texts and ancient modes of thought, will go hand in hand with Religion; and that, although theological control will continue to diminish, Religion, as seen in the recognition of a Power in the universe, not ourselves, which makes for righteousness, and in the love of God and of our neighbour, will steadily grow stronger and stronger, not only in the American institutions of learning but in the world at large. Thus may the declaration of Micah as to the requirements of Jehovah, the definition by St. James of "pure religion and undefiled," and, above all, the precepts and ideals of the blessed Founder of Christianity himself, be brought to bear more and more effectively on mankind."

And, 395 pages later, he concludes:

"Thus, at last, out of the old conception of our Bible as a collection of oracles, a mass of entangling utterances, fruitful in wrangling interpretations, which have given to the world long and weary ages of all hatred, malice, and all uncharitableness; of fetishism, subtlety, and pomp; of tyranny, bloodshed, and solemnly constituted imposture; of everything which the Lord Jesus Christ most abhorred has been gradually developed through the centuries, by the labours, sacrifices, and even the martyrdom of a long succession of men of God, the conception of it as a sacred literature a growth only possible under that divine light which the various orbs of science have done so much to bring into the mind and heart and soul of man's revelation, not of the Fall of Man, but of the Ascent of Man, an exposition, not of temporary dogmas and observances, but of the Eternal Law of Righteousness the one upward path for individuals and for nations. No longer an oracle, good for the lower orders to accept, but to be quietly sneered at by the enlightened, no longer a fetish, whose defenders must become persecutors or reconcilers, or apologists; but a more fruitful fact, which religion and science may accept as a source of strength to both."

In the middle of these discussions, with his tongue in his cheek, Arthur C. Clark let this little conundrum loose upon the world; which he titled *The Speed of God*, "He's coming just as quickly as He can."

"For some years I have been worried by the following astro-theological paradox. It is hard to believe that no one else has ever thought of it, yet I have never seen it discussed anywhere.

One of the most firmly established facts of modern physics, and the basis for Einstein's Theory of Relativity, is that the velocity of light is the speed limit of the material universe. No object, no signal, no influence can travel any faster than this. Please don't ask why this should be; the universe just happens to be built that way. Or so it seems at the moment. But light takes not millions, but billions, of years to cross even the part of Creation we can observe with our telescopes.

So, if God obeys the laws He apparently established, then at any given time He can have control over only an infinitesimal fraction of the universe. All Hell might (literally) be breaking loose ten light years away, which is a mere stone's throw in interstellar space, and the bad news would take at least ten years to reach Him. And then it would be another ten years, at least, before He could get there to do anything about it.

You may answer that this is terribly naive that God is already "everywhere". Perhaps so, but that really comes to the same thing as saying His thoughts, and His influence, can travel at an infinite velocity. And, in this case, the Einstein speed limit is not absolute; it can be broken.

The implications of this are profound. From the human viewpoint, it is no longer absurd—though it may be presumptuous to hope that we may one day have knowledge of the most distant parts of the universe. The snail's pace of the velocity of light need not be an eternal limitation, and the remotest galaxies may one day lie within our reach.

He is coming just as quickly as He can, but there's nothing that even He can do about that maddening 186,000 miles per second. It's anybody's guess whether He'll make it here in time."

Stephen Hawking devotes a whole chapter to the subject of God, and the creation of the Universe, in his book *Black Holes and Baby Universes and Other Essays*, plus many references to God throughout the book. In 1992, the BBC broadcast an interview with Stephen Hawking; it was conducted by Sue Lawley. This interview concludes his book—here is an excerpt:

SUE: To oversimplify your theories hugely, and I hope you'll forgive me for this, Stephen, you once believed, as I understand it, that there was a point of creation, a big bang, but you no longer believe that to be the case. You believe that there was no beginning and there is no end, that the universe is self-contained. Does that mean that there was no act of creation and therefore that there's no place for God?

STEPHEN: Yes, you have oversimplified. I still believe the universe has a beginning in real time, at a big bang. But there's another kind of time, imaginary time, at right angles to real time, in which the universe has no beginning or end. This would mean that the way the universe began would be determined by the laws of physics. One wouldn't have to say that God chose to set the universe going in some arbitrary way that we couldn't understand. It says nothing about whether or not God exists; just that He is not arbitrary.

SUE: But how, if there's a possibility that God doesn't exist, do you account for all those things that are beyond science: love, and the faith that people have had and have in you, and indeed in your own inspiration?

STEPHEN: Love, faith, and morality belong to a different category to physics. You cannot deduce how one should behave from the laws of physics. But one could hope that the logical thought that physics and mathematics involves would guide one also in one's moral behavior.

SUE: But I think that many people do feel you have effectively dispensed with God. Are you denying that, then?

STEPHEN: All that my work has shown is that you don't have to say that the way the universe began was the personal whim of God. But you still have the question: Why does the universe bother to exist? If you like, you can define God to be the answer to that question.

Personally, I have thought about this question for a long time; my doubts began in my late teens when I thought about such conundrums while riding the plow on our farm in Manitoba. I can summarize my thoughts this way:

Yes, the concept of a God exists, though in many different forms, depending on the culture in which you live. But that belief is in no way proof of His existence. In other words, God's existence resides only in the minds of mankind, not in some nebulous region of outer space. If God is considered in this context, the behaviour of religious societies begins to make some sense. It is axiomatic, that throughout human history, we can assume that once human beings developed a means of communication, they gazed upon the stars; thought about their own existence; worried about an after-life; pondered the fate of their souls—all this in whatever mode of expression was/is available to them.

Since the answers were not then (and are not now) available to mankind, people have constructed answers that provided some solace, namely myths and eventually, religions. Those myths became the means by which the shamans of ancient cultures derived their authority.

In a remarkable book, *Hamlet's Mill*, by Giorgio de Santillana and Hertha von Dechend, which they subtitled *An Essay on Myth & The Frame of Time*, the publisher's note on the dust cover, states in part:

> Contradicting many current notions about cultural evolution, this exploratory book investigates the origins of human knowledge in the archaic, preliterate world. Selecting Shakespeare's *Hamlet* as a congenial introductory figure, the authors

begin their journey proper with Amlodhi, Hamlet's counterpart in Scandinavian myth.

The mythical Amlodhi was the owner of a fabulous Mill which, in his day, ground peace and plenty. Later, in decaying times, it ground out salt. Now, at the bottom of the sea it grinds rock and sand, and has created a whirlpool, the Maelstrom, which leads to the land of the dead. The ultimate significance of this Mill, and of many similar mythical constructions, is what the authors set themselves to discover.

The trail, pursued necessarily by induction, leads around the world through many lands, Iceland, Norway, Finland, Italy, Persia, India, Mexico, and Greece, to mention only a few. It also recedes in time until the beginning reached several millennia ago in Mesopotamia.

As innumerable clues emerge and begin to interlock, several conclusions become inescapable. First, all the great myths of the world have a common origin. Next, the geography of myth is not that of the earth. The places referred to in myth are in the heavens and the actions are those of celestial bodies. Myth, in short, was a language for the perpetuation of a vast and complex body of astronomical knowledge.

The implication of these findings is no less startling for being self-evident. If, hundreds of centuries ago, man's mind could formulate a consistent and magnificently intricate cosmology, then clearly that mind had already transcended the influence of any evolutionary process. The authors say, along with the now forgotten Dupuis at the close of the eighteenth century: "Mythology is the work of science; science alone will explain it."

If you accept my argument so far, that God is a human concept and has no other existence, then questions such as, "Why does an external benevolent God allow so many injustices in our society?" become meaningless, for that God is within ourselves. Another outcome of this interpretation is that the concepts of heaven and hell have no validity. Regretfully, we will never see our loved ones in an after-life, therefore it behooves us to ensure that we behave in a loving and compassionate way while we are aware of ourselves on this particular planet. (A side issue that is receiving some attention, is this: are animals, particularly primates, and possibly elephants, aware of themselves as part of a society? If so, do they have a concept of mortality? It would appear that a fundamental difference between ourselves and other animals comes from the development of a civilization that has knowledge of its history.)

A further corollary is that God has nothing to do with the origin of the universe; that remains a scientific question, which may eventually be beyond our understanding. As Hawking says: "If you like, you can define God to be the answer to that question."

If our societal problems are viewed from this perspective, it should be possible to understand much of human conflict, not in a religious sense, where much of human conflict originates, but rather as a conflict of interests, namely territorial rights, be they rights of power or of territory. Since I have arrived at this position, I personally, have been much more at peace within myself as I see many of man's problems arising from believing the myths of his primitive beginnings and now his arrogant view of the world around him.

Summary

Du Noüy: The concept of God is necessary, therefore He exists.

Wright: God and Religion can co-exist with Science.

Hawking: There is no need to hypothesize God to explain the Universe—physics will do that.

Clarke: God cannot have a significant influence on mankind—He is elsewhere.

De Santillana: Myths and Religions have a similar, if not identical origin. Science will explain them.

McNarry: God does not exist. The human concept of God exists and consequently has a strong influence because of that belief.

*A concluding note: this essay arose in my mind as I watched a recent television broadcast of *Counter Spin*, conducted by Avi Lewis, in which the discussion was about Creationism and Darwinism. In the introduction, Avi Lewis mentioned Stockwell Day's comments recently in which he stated that he believed that mankind had existed for only 6000 years, and that mankind had walked the earth along with dinosaurs. Here, I show my biases when I say, No wonder we have political problems!

CHAPTER FIFTEEN
DOES GOD EXIST?—1988

The following is an excerpt from a letter I wrote in reply to the adult daughter of a long-standing friend of mine, who was a member of the Christian Science Church. She was remembering a conversation we had at their dinner table several years previously. In her letter she says: "*During a conversation I had today I was reminded of your promise to write me and tell me why physics will NOT prove the existence of God.*"

After some preliminary remarks—my reply: "Now to your letter.

May I say right off the top, that you will find many of my ideas shared by Carl Sagan in his book *BROCA'S BRAIN* (available as a paperback) in his chapter "A Sunday Sermon".

The question of a supreme being has occupied mankind ever since he has been able to ponder the significance of his existence and ask "Who am I?", "How is it that I am aware of my own self?" and "What happens to me when I die?" among others, in whatever language/s he/she/they use/s.

These questions are unanswerable. They also are at the core of the questions you are asking. When I say they are unanswerable, I mean that unambiguous answers that are provable to all people are not possible. Answers are provided in many different ways and contexts—most of them would be called religious. This of course, means that the answers are matters of belief and faith <u>not</u> physics. Physics may be defined as that body of knowledge that attempts to explain the physical universe around us in ways that allow for extrapolation to new, but contiguous experience. This definition excludes any explanation of what we have called spiritual experience. Spiritual experience therefore must be what we experience inside our heads as perceptions which have no physical counterpart in the so-called real world. This has been at the core of the deliberations of the great philosophers for centuries past; and will be for centuries to come.

Now, what do I think about all this?

Very briefly, it is my belief (here we go again!) that God exists only as a perception within human minds and has no other significance or existence. "He" or "She" exists only when and where we, that is people, think of "Her". By this definition heaven and hell do not exist either. If so, what happens when we die? George Wald (a Nobel Laureate) in a series of CBC lectures entitled "*Therefore Choose Life*" described life as existing as a bead on an infinite string or thread—the beads come and go, but are always part of an infinite existence.

The other concept is that we fade slowly after we die, because we continue to exist in the memories and perceptions of those who knew us and thus our influence continues in those perceptions, and in the things we physically made, the letters and books we wrote, and the many other by-products of our existence.

Religions are formalized ways of organizing our response to these questions. Unfortunately, (from my point of view) religions become ritualized applications of man's—and I mean the male of the species—hunger for power over others. Mary Baker Eddy is an exception. She had a fine idea but a very poor understanding of biology. Religions also assumed to serve a useful function in maintaining a certain moral code, with adaptations to particular groups of people.

I tend to think that our basic moral (as opposed to ethical) behaviour derives from our genetic heritage. Over millennia of time certain behaviour patterns of a co-operative nature were effective in preservation of the species and thus became embedded in the genetic code. With the development of civilizations this natural code became altruistic and was again encoded into religions, together with a ritualistic threat of damnation if ignored.

All of which is not to say that these behaviour patterns are wrong, but rather our perceptions of them come from a different source than the probable reality of our world (as I perceive it!).

CHAPTER SIXTEEN
THINKING, EMOTIONS
AND LOGIC—1989

How thinking and logic can over-ride our emotions and the consequences thereof; compassion and empathy are also involved. I don't remember much about Bloom's discourse, but I can recall instances when I was concerned about the lack of empathy in human relationships; when logic was used without compassion, with disastrous results.

TV ONTARIO broadcast a series about artificial intelligence and the lack of compassion and commitment in student attitudes and thinking, as seen by Allan Bloom, University of Chicago. He was interviewed by Robert Fulford. I recorded some of my reactions:

Most emotion is felt and reacted to, without a detailed rational thinking process. That is not to say that no thinking takes place; it does, but it may be limited and may not be at a highly conscious level. Bloom was concerned that students seemed not to be aware that they could think of themselves and their lives in an exciting futuristic way and in that process feel deep emotions. He seemed to be putting much of the blame on changing value systems within the universities.

Thinking involves the continuous association of "chunks" of previously learned information in new and different ways. The AI [artificial intelligence] concept of "if this then that" is a critical part of the process, although it will seldom be a conscious part of it.

Growing up in close contact with the natural world—that is the world as it has evolved in the part of it where we came to have our being— imprints images that we use in our thinking processes. If those images do not involve the "natural" order of growth, aging and decay followed by rebirth in the endless cycle of the natural world, then our sense of time and the cycle of support that permeates the natural world will be primitive or at least limited. Instead, we may well have a sense of the "natural" order of things that is literally related to non-living things. Our thinking then will be lacking in compassion, except as it relates to our own sense of physical well being and survival.

Then we walk by on the other side of the street while the elderly lady is robbed, or the inferior person of another race is shot, or a whole nation is starving—they are linked to us through the mechanical (electrical) mode by which we developed our thinking processes.

EMOTIONS vs. LOGIC

Most of us have, at one time or another, wondered about our emotions. They seem to arise unbidden from some unknown source; our hearts beat faster (affairs of the heart!); we react with trepidation or elation; most attempts to analyze them fail in the face of logic.

We, of course, are biological beings who function according to our biological natures. Most of us watch TV with no knowledge of its interior circuitry; to us, all TVs operate by being plugged into a power source (we don't know where that comes from either!) and we see a picture on the tube and hear related sound from a speaker. In fact, each TV operates according to well defined principles of physics, just as we operate by well defined principles of biology.

But the emotional content of the images we see is only possible because of the fundamental principles operating in both the TV and ourselves.

Those of us who have some understanding of basic scientific principles also experience emotions, sometimes they are overpowering, but experience tells us that they can be controlled, even if we cannot explain the logic of doing so.

On the other hand, for those of us who (and we are in the majority) experience our emotions and react to them from a background of real experience, but without any knowledge of the detailed principles involved, our emotions thus become the reality from which we react.

In an evolutionary sense, it is likely that emotions arose from the need to flee instantly from danger, rather than to stop and analyze the situation. If we succeed in surviving, it is afterward that the analysis may be useful. Over millennia of time, the emotional responses have become very refined and the danger signals are often triggered by what is perceived as an assault on our ego or psyche, rather than a danger to our physical person. Our psyche is the result of our life experiences, which are, of course, different for each one of us; we all have a psychological part of our being, whether we recognize it or not and we all have our own psyche. We must realize that we do not have to be a slave to either our emotions or an impersonal logic to explain them. What is important is that we do not let either emotions or logic rule our lives.

We as human beings have created wondrous and emotionally satisfying works in the form of music, art, architecture and an understanding of the world around us; a scientist's highest praise for a theory is that it is elegant; a love poem sends us into raptures of fantasy.

Not all human activities are beneficial; men are not gods; they are fallible, both in their creations and their understanding of them. It is in understanding the relation of our emotions and the use of power in human society that we have the least understanding. As we grow, this understanding changes, as it should, and our perceptions also change. This is completely natural and should not be resisted; if it feels good it probably is; if our feelings distress us then maybe we should try to selectively react to those that feel good; we all grow and change and that too is one of the glories of being human.

CHAPTER SEVENTEEN
CONCERNS ABOUT SCIENCE, TECHNOLOGY AND MODERN SOCIETY—1990

A letter to Peter Gzowski, (CBC) November 1, 1990

Dear Peter:

I have just listened to your interview with Alan Gregg about his new book on the Canadian mood.

Your last question about "How do the citizens of this country identify themselves" (paraphrased) reminds me of a conversation I had at a banquet table in Winnipeg about 20 years ago. The discussion centred around the same topic; I said that my sense of who I was derived from my identification with mankind in general, i.e. the inhabitants of this earth and then with the country known as Canada. I was ostracized for the rest of the conversation!

Gregg's book and the recent CBC/*Globe & Mail* poll plus the recent Ontario election all point to a very deep unease, not only in Canada, but in the industrialized world. The vast and unprecedented changes in Europe also point to a fundamental restructuring of the organization of societies.

Let me just say that I am being forced to the conclusion that some time during the next century we will be forced to abandon the free-enterprise-capitalistic notion that each of us is free to pursue what we want with little regard for the rest of humanity.

It seems inevitable that what must come is a vast reorganization that operates on the realization that the prehistoric open-fire hearth now has become a global hearth and that we all must put our best into the same pot and from that pot comes our basic survival needs. What we do as individuals after those needs are met is to a large extent dependent upon our location on the planet and the interactions with our immediate neighbours.

Idealistic? Perhaps, but I see no other way except the horror of terrible destruction and the rebuilding afterwards. It is this kind of vision that our present system of governance, with its adversarial biases, fails to contemplate.

Keep on prodding us!!

Bob McNarry

THE BEGINNINGS

Most of us accept the concept of evolution of life on this planet. This means that in the beginning there was no science, no technology and there were no politicians.

The earliest societies had only their native intelligence, their language, their memories and their stories to use in understanding, adapting and using the available products of the natural world about them. Gradually, methods of encoding knowledge were evolved and the fund of collected experience became larger and more available so that civilizations (societies that developed a sophisticated culture) evolved. When science, the encoded knowledge of the natural world, became a part of these civilizations, its application to human needs and perceptions resulted in the development of industry and with the further development of engineering tools, modern technologies arose.

TECHNOLOGY, POLITICS AND CHANGE

Along with these developments, societies evolved ways of formalizing their civic functions; and political systems also arose, along with nation states. The people living through all these vast changes were undoubtedly unaware, on the larger global scale, of the changes themselves or of the eventual global effects.

Today, terms such as environment, pollution, poverty, war, annihilation, trade wars, tariffs, political oppression, loss of fundamental human rights etc. all remind us that technological change has indeed altered our lives. The terms just listed are all negative in connotation.

What is it, then, that has produced such widespread negative effects? Surely it is not the natural world itself; it is not the animals of this world, (excluding homo sapiens). Rather it is how we, as people, have structured our own societal behaviour and used our technological knowledge. And what is the engine that drives the use of technology? It is an economic engine. Our societies evolved around a hierarchy of power and authority. Our interactions as groups have evolved around trade and from that grew economics as a way of understanding and managing trade. Almost all trade is based upon one of the trading partners gaining an advantage. We now measure that advantage by the size of the bank account; but where is the concern for humanity in that bank account?

Let us consider what we do when we educate our children, and by implication, ourselves.

We teach our children something of our culture, our history, how to be numerate and how to be healthy. We hope they acquire a feeling for the value system under which we live, that they have a sense of national pride, that they can use their natural talents in a satisfying, productive way. But do we not also isolate them from any real perception of the effects of our technology on the natural world? If we have taught them about these effects, how is it that we have so much pollution, war and destruction? What happens to our children as they progress through adolescence to reach adulthood and responsibility?

Do we not make decisions based upon immediate expediency rather than long term understanding and a sense of trusteeship for the future? Is not our political system devised so that our attention is focused on the next few years rather than how our actions will affect those living in the next century?

You may ask—but how can we possibly see that far into the future when we have difficulty seeing ten years ahead? The answer is not clear, but what is becoming clearer is that we must try. The combined interactions of our technological, economic, trading, agricultural, communication, political and societal activities is now so complex that we can hold a two-day meeting of First Ministers to discuss some of these questions and then not even agree what the questions are, let alone find any solutions.

To go back to the question about engines. The engine is made up of several parts. It is an engine that, in the main, is made from technological parts, is governed by agreed upon procedures, and is fuelled by the wealth extracted from all members of our society. However, the fuel is not extracted equally nor is the output evenly distributed. Perhaps it can never be equitable in operation. The forgotten parts of the above equation are "Where does the energy come from?" and "What do we do with the exhaust of the engine?".

Fred Hoyle, an astronomer, once remarked that any civilization arising on any planet of any star in the universe has to make the transition to the use of nuclear/stellar (solar) energy using the available natural energy sources of that planet, otherwise that civilization will not survive. What he is saying, is that we must find alternative energy sources to fuel our technologies before we burn up all our petroleum, coal and wood. If the Egyptians had developed the equivalent of our energy consuming technologies 3000 years ago where would we be? We have very little time left on an evolutionary scale.

Moreover, one of the "exhausts" of our "engine" is pollution, unnatural wastes that come from chemical combinations that are by-products of our technologies. Don't forget that even solar energy is not pollution-free if it is used to drive the industrial "engine".

What all this is leading up to, is that our present political system of selecting people from our population and giving them authority to govern our society has not been able to effectively manage either our "hearth" or our environment. It is quite conceivable that the "cave man" going out to search for the next meal could read and understand his environment with more insight than we can. At least he seemed to have respect for it and understood that its nurturing quality depended upon its preservation (ask our North American Indian friends). We, in our turn, seem to regard the bank account as the great "mother" from whom all blessings flow.

WHAT CAN WE DO?

We can re-examine our political system and try to evolve one that ensures that there are representatives in our governments who are aware of the whole of the complexity that permeates our culture; who do not take precipitate action to satisfy political whim or dogma; who understand that there are long term consequences to any action taken today; who can discriminate among different scenarios; who can articulate and explain what they understand; who can be persuaded to act for necessity rather than expediency. Otherwise the noble experiment we have called civilization may be just that—an experiment.

This is a heavy burden. It won't be resolved quickly or easily. We must make it central to all our considerations. It will not come about because of a few passionate writings. We must not be afraid to face the future and we must try to prepare, as best we can, for the inevitable consequences of squandering our heritage. We must assume our full responsibilities to the unborn generations yet to come.

Babies are innocent—we are guilty.

There are, in any population, those whose lives are dedicated to questing for knowledge. They do not seek public adoration, but they do want their findings to be respected and considered. They are like the scout on the hilltop seeking the way through to the next valley and over the next mountain range. Without them we are lost, or at best, forced to wander on our valley floor forever seeking an exit, until we die exhausted in the attempt.

It is for the politicians to understand and interpret the scout's information so that the way into the future is negotiable with a concern for the quality of life for everyone. A noble and responsible task.

CHAPTER EIGHTEEN
THE CANADIAN DILEMMA—
MEECH LAKE—1990

My reactions to the CBC broadcasts of the Constitution debates and the Meech Lake dilemma were strong and visceral.

It would seem that the mood of the country is volatile and that there is a strong sentiment for regional control. Probably due to the perceived failure of the Federal Government (generic) to adequately understand and deal with a changing country and world.

We seem to be realizing that Constitutional entrenchment of rights and relationships lead to frustrations and excessive legal fees. Perhaps the British tradition of an "unwritten constitution" might be better. Women are now realizing that they do not have as much protection as they once thought after the passage of The Bill of Rights & Freedoms.

ITEM: Bourassa is a technocrat and therefore the comment that he did not realize the consequences of Bill 178 is valid. His objection to changes in Meech Lake stem from his obsession with the power resources of the James Bay area.

ITEM: Newfoundlanders are still angry about the Smallwood hydro deal with Quebec (and they should be!) Bourassa is now trying to get more power because they have oversold to the USA. Statesmanship loses to power politics!

ITEM: Lougheed had the same objection re Alberta's resources. Don Getty has no ideas! I doubt if he even understands his own position on Senate reform. He certainly does not understand that it will be of little or no help to Alberta.

ITEM: Peterson's appearance before the Meech Lake Committee was more important for what he did NOT say than for what he did say. He laid down a smoke screen of concern for the country, but did not address specific interests or concerns or the content of the Accord. He was one of the last hold-outs at the Langevin Block meeting re senate reform.

ITEM: Discontent with the current Federal Government is widespread. People are talking about the need for revolt. While not a current possibility, the sentiment is there and could grow if the economic and cultural situation were to significantly worsen.

ITEM: The eventual union of Mexico, USA and Canada is probably inevitable. Europe is setting the stage. Events and situations evolve at a much more rapid pace than most of us realize. The rate of change can only increase with advancing communications and information technology. What is important is to try to preserve as much as possible of our culture and institutions during these changes. Meech Lake will be seen as a blip on the screen of history.

ITEM: Mulroney has no sense of the mood of the country; remember Peter Gzowski's last question during their interview. He fudged his answer. His personal thinking was revealed by his comment in Calgary re two dissident MPs, "Millions voted for me!"

Peter Gzowski hosted a Morningside discussion on Meech Lake; Senate reform; and the general sense of crisis in Canada, June 1, 1990.

They all missed the real point! The real point is that Canada now has the opportunity to adapt its governmental system to the requirements of the next century! Why can we not see this?

Because, like all of mankind's organizations, current governments are all about gaining and keeping power and control; the electorate are deluded in thinking they are merely giving their authority to governments for a short time to do the electorate's bidding—a snare and delusion! The church and the feudal system used to exercise that power, now it is the politicians, and even they are at least partially under the fiscal control of the multinationals. We, the people, can regain it (maybe we never had it!) by reorganizing the political map from provinces and geographic regions to societal functions, thus redistributing those powers in a different way.

Societies will always, because we are human, be power based. The fighting between the males of the animal kingdom for the right to breed is universal and this includes mankind. The expression of that fundamental evolutionary factor may differ, but the result is the same; the most powerful rule. Where we differ, is that we have a sense of history and culture. We are not unique in our sense of compassion, which is also shown by some animal species.

The entire discussion this morning was extremely depressing because there was no hope in it; we need hope above all else, if we as a country are to survive. It is hope that makes life worth living. (See my book, *The Sacred Fire*.)

If Mr. Mulroney had taken a nation-wide vision of this country to that evening in the Langevin Block in Ottawa and used it, rather than using his labour negotiating tactics as a basis for developing the agreement, we would not be in this crisis situation today. Sure, it was "Quebec's round", but he was so anxious to go down in history as the great conciliator that he sold his soul to the other premiers and gave them as much of what they wanted as was necessary to get a deal. Mr. Bourassa agreed because, having their signatures on a piece of paper, he fully believed that the Accord would be ratified within the year and then he was "home free."

Mr. Mulroney, when he was seeking the PC leadership, condemned anyone who would even consider taking "separatists" into his government. Yet what did he do? He was so anxious to get elected that he did just that, on the pretext that they would become converted federalists—did they?

We have, given the situation in Europe and the Soviet Union, a heaven-sent opportunity in Canada to show the way for the inevitable global integration of all societies. It should allow for cultural enclaves to evolve, freely organized, but subject to the necessities of global economic interactions. I know that my earlier remarks about power would tend to defeat any such arrangement, but perhaps there is enough hope and tolerance in our humanity for it to succeed. After all, isn't that the message most Christians in this country espouse?

I have been so frustrated by the discussions in Parliament, on radio and TV; almost all were so self serving—using innuendo, omission, distortion—that no clear understanding was possible. I have had an uncomfortable feeling for some time now that the CBC TV coverage had a hidden agenda. The leaked Ontario document did nothing to dispel that concern. The "glamorous" coverage by *The Journal* documentaries was revolting; the subject is far too serious for that kind of presentation. The best coverage, and by far the most useful, was from the Charest Committee hearings. We at least had options to consider.

Here's to Mr. Clyde Wells! On radio and TV talk shows he clearly is the nation's hero in this debate outside of Quebec—though I have heard callers from within Quebec praise him. Recently, I have noticed that more callers are saying they wished he was our Prime Minister. One may not agree with him, but you know exactly where he stands, and that I prefer to Mulroney's deviousness. There is more than a suspicion that Mr. Mulroney is afraid of him.

SENATE REFORM

The appointed Canadian Senate has, for many years, been regarded as an anachronism in our parliamentary system. The NDP has long called for its abolition. Currently, its reform is one of the most divisive factors in the Meech Lake debate.

Western and Atlantic Premiers want a Triple E Senate—Elected, Equal and Effective. Western Canada wants more input into the Parliamentary system. The Atlantic provinces also feel alienated; while central Canada, Ontario and Quebec, are satisfied with the current arrangement. They both want veto powers over any modification or reform. This situation leads to a classical stalemate.

As we enter the 21st century, should we not consider development of a governmental body to replace the Senate, which will be truly effective in a New Age—the age of computers, information management and global ecological concerns?

The current system, based on antagonistic political parties, will continue to waste valuable time and resources, while global problems all about us increase at a dramatic rate. Let us therefore consider an entirely new system for Canada.

As a first step, let us consider developing an elected Senate (a different name is essential and will inevitably be devised) which is not based upon Party or Provincial affiliations, but rather on a realistic geographic and economic representation of the concerns and activities of the Canadian people.

To do this will require careful crafting and much discussion, along with a great deal of tolerant optimism.. What I have in mind is a Senate elected, not from the provinces, but from the economic sectors of the country. The difficulty will be in defining these sectors and working out the powers of the new body.

One immediate consequence would be committed, knowledgeable people representing the agricultural communities, the fishing industries, the forest industries, the financial community, the health care community, the manufacturing community and so on.

The physical facilities are already in place in Ottawa, the necessary support infrastructure could be adapted from the existing bureaucracy; the election practices would have to be modified somewhat, and the relationships with the House of Commons would have to be worked out. In all probability the Senate could craft bills and legislation for Commons examination and final enactment into law. In a sense, this would reverse the roles of the two bodies now making up our Parliamentary system.

Think about it.

* * * * *

MEECH LAKE

Some comments in reaction to a local newspaper article, which posed these questions:

1. Why does Meech Lake dwell on Quebec?

We all know that Quebec has not signed its acceptance of the repatriated Constitution because its uniqueness and distinctness has not been recognized with an increase in its powers to control its own destiny, and that Meech Lake was designed to correct this situation and so permit Quebec to be a signatory of the Constitution.

2. Was Meech Lake dreamed up in a locked room on the spur of the moment?

This is a deliberately confused question. The final agreement on the Meech Lake accord *was prepared in a room, late at night, without allowing access by the participants to their constitutional advisors.* There were lengthy meetings and constitutional debates over a period of years in preparation for this meeting. But the final agreement was hammered out, and agreed to, in a closed room under considerable pressure.

3. But wasn't there something undemocratic about the whole thing?

The undemocratic thing was not that there were no lengthy discussions preceding the wording of the accord, but that there was no subsequent discussion on the actual wording of the Accord to determine if it did truly represent the intent of the previous deliberations. The precedent of the "kitchen cabinet" meetings which proposed a solution to the 1982 Constitution without Quebec's input is not a sufficient reason to say that Meech Lake was a more democratic process. Neither was good constitutional practice.

4. But didn't we do it differently in 1981-1982?

How can we say that Meech Lake met a higher standard than the 1982 constitutional changes, when both were flawed?

5. But can't we make the constitutional reform process more democratic in the future?

Of course we can! But to say that we must accept an admittedly flawed document because it has some benefits is an unworthy argument for a mature nation.

6. Why can't we change the accord right now, as many critics demand?

The argument to pass Meech Lake now and patch up the other constitutional concerns later, particularly in view of the "unanimity" clause smacks of desperation to achieve a "behind closed doors" agreement. The constitutional uncertainty as a result of passing Meech Lake will most assuredly be damaging to all Canadians.

7. What does the accord say?

The discussion of this question purports to discuss the "distinct society" clause. It neatly leaves out any discussion of the legislative "role" of the Government of Quebec by not quoting the entire clause.

8. Why is the distinct society clause so important to Quebec?

In general there is no reason to dispute this question, except for the use of the word "self-confidence"; Quebec citizens probably have more sense of identity and self-confidence than most Canadians.

9. Does this mean "special status" for Quebec?

This analysis seems to be a lengthy circumlocution of the meaning of Section 2(3) without an indication that there are not any restrictions of the "role" of the Quebec legislature. In other words, the Quebec legislature can enact any law under the guise of "distinctness", subject only to the interpretation of the Supreme Court of Canada. These powers are not available to any other province. If such powers are to be given to Quebec then they must be specific and not generalized.

10. But doesn't the distinct society undermine rights enshrined in the Charter of Rights and Freedoms?

Agreed; the Charter of Rights and Freedoms will prevail, under the interpretation of the Supreme Court of Canada. But can we be assured that provincial rights, assumed under the Meech Lake agreement, will not become, through validity of established use by being unchallenged, entrenched as rights and then we would have not only distinctness, but inequality of citizens within Canada?

Many Quebec citizens now feel that they suffer inequality, and if they perceive themselves as suffering inequality, they in fact may be disadvantaged. Meech Lake is said to provide that sense of equality while allowing distinctness, but it imposes few bounds and could in the future result in constitutional and legalistic disputes.

If Quebec would renounce forever any attempt to disassociate itself from Canada then perhaps an agreement might be more easily reached. Of course they would lose their most powerful bargaining chip. But then perhaps statesmanship, rather than labour bargaining tactics might be more appropriate.

The major problem with the relationship of Quebec and the rest of Canada is that governments have, over many years, tried to impose cultural change on Canadian society by legislative fiat. (Don't shove French down my throat!) We have only to look at Europe to see that it does not work. Societies eventually work out their own adjustments. The role of governments should be to facilitate this process without the use of violence or economic warfare.

11. What about the rights of linguistic minorities: English in Quebec and French in the rest of Canada?

The rights of linguistic minorities have had a varied history in Canada. Manitoba passed its discriminatory schools legislation, and in general the use of French in Western Canada has not been given much support. In Quebec, the English minority has been given excellent facilities to continue operating in its own language. In large part, this has been possible because of their dominance in the financial and business life of the province. With the increase in francophone participation this has changed. While not physically displacing the English component of their society they have assumed increasing control and, perhaps not wisely, but certainly not surprisingly, they have passed legislation that facilitates the use of French in all aspects of life in Quebec. Herein lies the heart of the problem.

What a future Meech Lake must do, is to establish conditions which allow this adjustment to proceed under the wisdom of society, not legislative decree.

We must clearly distinguish between negotiations and statements by governments and the perceptions of the population of any Province. Referenda and elections make those distinctions.

12. What about women's rights?

Why even allow a situation where women's rights may need any protection against another possible source of discrimination?

13. Doesn't the accord undermine the rights of Aboriginal people?

This is such a fundamental injustice that the whole of Canadian legislative action stands indicted.

The inclusion of the "unanimity" clause will make any resolution of native rights extremely difficult, if not impossible, to solve. Again, Meech Lake is severely flawed.

14. Doesn't the accord weaken the powers of the federal government?

The main problem for the federal government is the inclusion of the "unanimity" clause. It will be very difficult to make future changes without concessions to the dissenting provincial governments. In this sense the federal government will be at the mercy of the provinces. The word "Balkanization" is appropriate.

TANKS
There were no tanks in the street today,
Nor yesterday,
Nor the year before that,
No, there were no tanks in our street today.

But there were tanks in the streets of Gdansk,
And in Budapest,
And in the streets of Kabul,
But there were no tanks in our street today
—Yet.

Written after contemplating the pictures
in Lech Walesa's autobiography.

January 28, 1989 [Note: Written before OKA!]

Subsequently I , with help from my daughter Margaret, prepared a draft of a new mode of governance based upon the economic, geographic and cultural nature of our country. It used a voucher system to finance elections from public funds. I sent a copy to the committee charged with collating suggested revisions of the Constitution; The Honourable Joe Clark responded with a three-page letter saying my plan was unworkable. He may have been right! I wanted to stimulate some new thinking as we entered the 21st Century.

CHAPTER NINETEEN
ON THE CBC BUDGET
CUTS—1990

Politicians do watch, and listen to, the CBC; they don't like what they see or hear. They don't seem to realize that the CBC policy is to try to give all sides a chance to comment; too often the government response is "No comment." Invariably, the government's fiscal response is to cut the funding for the CBC. Currently [2012], Prime Minister Stephen Harper wants to abolish the CBC entirely. Except for my stint overseas during WWII, I have listened to the CBC since its inception in 1937. I began recording CBC broadcasts in 1958—eventually stopping in 1992. The National Archives copied about 700 hours from the early tapes. That the CBC tried to present a balanced view of Canadian politics, regardless of the political party in power, is clearly evident.

The cuts announced today, December 5, 1990, by the CBC Management are the direct result of the Federal Government's fiscal and budgetary policies and the "Made in Canada" recession. This is not to say that the CBC Management is beyond criticism; one of the problems with any large bureaucracy is the inevitable "empire building" that goes on at the expense of the original purpose of the "corporation"; in the case of the CBC, this is programming: local, regional and national.

It is well known that Conservative governments are more disposed to favour private broadcasting and have little sympathy for the CBC. It can be assumed, with some degree of assurance that this antipathy stems from a perception that the CBC is a left-wing organization at the programming level and that it is constantly sniping at Conservative governments, both provincial and national.

I have a collection of over 8000 hours of CBC program sound tracks on magnetic tape covering a span of over 30 years. Upon listening to this material while developing a database catalogue it is abundantly clear that this perception is incorrect. What the CBC has done over the years is to raise issues, discuss them from many (and often conflicting) points of view so that the listening (and viewing) public can better judge events within our own culture as well as those beyond our own boundaries.

It would seem that the present Conservative Government members gather much of their opinions about the CBC from TV news programs and commentaries dealing especially with the political life of Canada. By the very nature of their occupation as Members of Parliament, they have little time (and perhaps, inclination) to do otherwise. CBC radio is probably perceived as a lesser threat than TV to Conservative policies and governments, since it would seem that a smaller number of the public obtain their information, on which they base their political choices, from radio.

A poorly-informed public will be more docile and more likely to accept information fed to them through the filter of private business oriented TV and radio without critical questioning. If the current government's advertising budget were slashed to 10 per cent of its present amount, and that amount fed into the CBC budget, the CBC could probably carry on and the public would certainly be better informed and able to make value judgments more in keeping with the needs and demands of our entry into the 21st century. The Federally operated book stores around the country have long ago been closed; they were a source of essential information on many aspects of Canadian life. The material was at least, partially paid for by charging nominal prices for the material available.

It is obvious that governments, in general, are failing to meet the demands for information in spite of "Freedom of Information" legislation. Rather than fostering an informed public, they tend to regard such a public as a potential threat; the general political unrest around the world attests to this situation; it is not peculiar to Canada.

The election of the NDP government in Ontario is testimony to the dissatisfaction of the public with current governments. The mood of the electorate in the USA is also critical of all levels of government. More and more, one hears comments critical of the governmental "process" as well as of particular legislatures. It is becoming increasingly clear that a different selection process and a different legislative process are necessary if we are to survive into the next century without widespread anarchy in the western world.

It is also axiomatic that no government wants to be the engine of its own demise. Yet, it would seem that change is both essential as well as necessary, if we are to learn anything from the events that have changed the face of Europe. The current Canadian practice of demanding party unity and of governing with an adversarial process in our legislatures has long since become irrelevant in understanding and devising policies that will guide us into and through the technological and information driven 21st century.

It is through intelligent questioning of our processes and values that we will manage our passage into the new Century. Agencies such as the CBC are essential in that process. The sensitivities of politicians in guarding their political turf will not, and cannot, provide that guidance. Certainly they will guard their "sacred turf" with all the guile they can muster, but that is not what we need.

Revolution will certainly produce change, but at unnecessary cost, both in lives and institutions, to say little of the material cost of property destruction.

The CBC is only one of many agencies that are useful in providing the essential input from which the citizenry must derive their knowledge and intuitions about their progress into the future. Private broadcasting is just as essential; we must all know all sides of any question. Newspapers, books, radio, TV and one-on-one discussions are all part of this process. But the essential difference in the role of the CBC is the one that Bill C40 is designed to eliminate, namely the role of fostering Canadian Unity. Of course many other organizations foster Canadian Unity, our public libraries are a prime example. But the CBC is the only fully national organization which can span this sprawling country with an immediacy and relevance that transcends all boundaries of geography, ideology, and provincial "inward looking" that tends to stifle our sense of national identity.

SOME FURTHER THOUGHTS ON
THE CBC BUDGET CUTS

The present Conservative Government has again displayed some of its inner thought processes by the manner in which it has mandated the reduction in the CBC's operating budget.

Earlier in 1990, the president of the National Research Council announced a massive reduction in the staff of Canada's only "National" research facility. President Pierre Perron is an appointee of the present Conservative government. The reason given is that the NRC must reduce its budget if it is to operate within the financial restraints imposed by the Federal Government. Again, the argument used is that ALL sectors of the Federal Government must be prepared to bear the burden of reducing the exploding national deficit. It is not without its irony that the President of NRC is referred to as "The smiling executioner."

Now think back to yesterday, as the President of the CBC, Mr. Veilleux announced his solution to the CBC's budget problems. He also is a Conservative government appointee; he also is following the Government's policy of making our National institutions *better* organizations by imposing fiscal restraint.

The recent cuts to VIA Rail were made and defended from the same premise. The Government said "Use it or lose it." Yet people had to book space, on the transcontinental run, months in advance if they wanted seats; it was a proven tourist attraction; but it was decimated in spite of the anguished outcry from Canadians.

Few things are more personal than a letter from a loved one; much of the lifeline of business depends upon the post office. Both of these functions truly transcend the boundaries of place and government ideology. Yet, our government wants us to commit these national functions to profit making privatization.

Air Canada has had a proud and enviable history in providing a national air service for this country. Yet, deregulation, which has caused much soul searching among those who care about public service and safety, has been used to weaken air travel safety and stan-dards throughout North America. Air Canada is now just another carrier, albeit with a proud heritage.

The National Research Council, The Canadian Broadcasting Corporation, Via Rail, the Post Office, Air Canada—all creations of our national Governments through the years, which have helped forge the bonds of National Unity across a very difficult land mass and an often hostile climate—all these agencies are now either gone or are in serious trouble.

It can be argued that most Canadians don't care one way or the other as long as their own personal situation is not materially affected. The Government response is always "Trust us to institute policies that will ensure economic growth and we will all be better off." Why then, are so many people so unhappy with this Government? Is it perhaps because "We do not live by bread alone"? Is not the soul of our nation in anguish? Where can we turn for those values we are taught to rely on in adversity? What has set region against region; culture against culture?

If we cannot communicate across our nation and interact with the different aspects of our culture; if we cannot utilize our best talents to understand and explore our natural world about us; if we cannot travel to see and experience our beautiful land; if we do not trust our mail to be delivered reliably and on time; if we do not feel safe flying across our vast land; how then can we expect to survive as a compassionate and considerate member of the community of nations?

I remember the beginnings of the Canadian Radio Commission, the forerunner of the CBC; I remember the thrill of watching the CBC grow as it provided a sense of national identity for an emerging nation; I remember my feelings of pride when I returned after serving in WWII to find an even more mature broadcasting agency; I remember watching the first Trans Canada mail plane fly across the wide prairie skies; I remember travelling across the rocky backbone of the Laurentian Shield and marveling at the skills and endurance of the builders of our railway; I remember my experience as a member of Canada's outstanding research organization, The National Research Council; I remember the absolute assurance that a letter placed in the hands of Canada's Post Office would reach its destination safely and on time.

I remember all these things, for they are part of the mosaic that made me a Canadian. They are more important to me than a slick new car; more and more violent TV to watch; more and more environmentally destructive commercial successes; more and more financially ruinous corporate mergers and buy-outs; fewer and fewer creations of human genius in the arts and culture, drowned out by physiologically destructive electronic cacophony.

My soul is in anguish and torment for my country. I do not know where to turn. Yet I must not give up all hope and turn to outright anarchy or revolution as a last resort. There has to be trust and hope in the Canadian people. My legacy for my children and yours, will I trust, be one that leads us back some small way to that compassionate and caring country I knew as Canada.

CHAPTER TWENTY
SOME THOUGHTS FOLLOWING THE CONSTITUTIONAL CONFERENCES—1992

The Honourable Joe Clark hosted a series of Constitutional Conferences across Canada, which were broadcast in their entirety. I followed them avidly and recorded many sessions on VHS tapes.

GENERAL IMPRESSIONS

The strongest single overall impression is of the wonderfully articulate "ordinary" Canadians who participated; they were far from ordinary! Their intelligence and eloquence was, in a sense, surprising; surprising because we tend to think of ourselves as rather stodgy and ordinary. We are articulate, compassionate, intelligent, and we love our country with fervour and passion.

There is, however, another impression that disturbs me greatly. It is related to the unease we, as Canadians, have when we consider our political system and our Governments. We are never sure which political agenda the politicians are using during any political debate.

During the Montreal Conference, it was alleged that the economic agenda was "hijacked" by the social activists. During the Vancouver

Conference, it was alleged that the "consolidation" agenda was "hijacked" by the business interests.

Both interactions during the conference were related to Proposal 14 of the Government constitutional document: "Broadening s.121, the common market clause." Proposal 14 is designed to facilitate the "free flow of persons, goods, services and capital" throughout the country. Associated with it is Proposal 15 that is designed to give the Federal Government sweeping powers to "manage" the economy.

Proposal 15 was effectively removed from the Constitutional agenda at the Montreal meeting; Proposal 14 was vigorously defended by the business interests, but was nevertheless relegated to the status of a political concern between Provinces and the Federal Government, rather than an entrenched constitutional matter, while the concept of a Social Charter was effectively placed on the Conference agenda.

It was during this time that the BCNI (Business Council on National Issues) appeared before the Dobbie-Beaudoin Committee to make their case for Proposals 14 and 15. They reappeared at the Vancouver Conference when Tom D'Aquino, as a workshop participant, again made a strong case for management of the economy. It is not known if he had already been selected as a participant, or if this was a last minute arrangement, designed to let the BCNI plead their case.

The Vancouver Conference was to be a wrap-up session where the decisions and consensus of the previous four conferences were to be consolidated as a guide to the Dobbie-Beaudoin Committee. This did not happen.

The division of powers, supposedly settled at the Halifax Conference again came to the fore and the common market clause was placed back on the table. The Triple E senate proponents did not accept the outcome of the Calgary Conference of Elected, Effective but Equitable rather than Equal.

What this means is that the expressions of goodwill and accommodation by the "ordinary" Canadians were fine, as long as the basic agendas of the special interest groups were not severely tampered with or modified. The "special interest" groups in this context are those that represent the major power brokers in the country, especially those of the business and political interests.

There were many other "special interest" groups at the conferences. Some were well organized, especially some of those representing women's interests and the Aboriginal peoples. The Social Charter issue was effectively placed on the agenda by the women's groups and the Action Canada group. It would appear that it was this tactic that first raised the hackles of the business interests when they complained that the Montreal Conference was "hijacked."

It was at the Toronto conference where the native issue of "the inherent right to self government" was also successfully placed on the agenda. The sense of accumulated guilt of Canadians forced the participants (with a large measure of genuine good will) to deal with this as a valid constitutional matter that must be settled during this round of Constitutional debate. The performance of the Aboriginal representatives throughout the series of five conferences was most impressive.

Throughout the five conferences there was an enormous amount of goodwill expressed towards the special needs of Quebec, but, in the end, the problems of devolution of powers to the Provinces remained a stumbling block; those who wanted a Triple-E Senate did not want any trade-off for special status for Quebec. The final outcome of this thorny problem is most unclear.

THE ECONOMIC ISSUE

The short, sharp exchange between Shirley Carr and Peter White in the televised Vancouver workshop session highlighted the deep-set suspicion of the BCNI agenda and their role in the whole constitutional debate. While never openly stated, it was apparent from comments made during each televised portion of all the conferences that the "social" special interest groups did not trust the Government's stated reasons for wanting to entrench economic control in the Constitution. The appearance of Tom D'Aquino heightened those concerns.

His main argument was that wasteful inter-provincial trade barriers, accumulation of massive government deficits, the need to become globally competitive, placed Canada's economic future in grave jeopardy. He further argued that economic good health and social programs are inextricably linked; we cannot have good social safety nets if we cannot pay for them. We must be able to create economic wealth, if we are ever to reduce unemployment and be an effective player on the world stage.

Those on the other side of the argument, claim that it is a fundamental tenet of Canadian society to ensure that all citizens have a right to some fundamental social services and support; that the economic players in this country have an obligation to support these rights in a real and meaningful way. To give them carte-blanche by entrenching control over the "movement of persons, goods, services and capital" in the constitution, would give them power to play the game by their own rules with social needs tagging along and getting what is left after the "business" needs are satisfied.

How the Government deals with these issues and concerns in their final drafting of the new constitution, will be a signal to Canadians, of the sincerity of the Government's declared intention to listen to the concerns of those extraordinary Canadians who participated in the five Constitutional Conferences.

SOME PERSONAL OBSERVATIONS

This morning, Roberta Bondar described her feelings and emotions as she looked down on Canada from the shuttle Discovery.

Constitutional squabbles become insignificant from that perspective; they also become insignificant when viewed from the long perspective of the evolution of mankind on this planet and our emergence into a dimly viewed future.

We still are little more than primitive tribal societies in the manner in which we exercise control over each other in economic and political matters. Economics has no natural counterpart in the real physical or biological world—it is a purely human construct. Politics, on the other hand, is a direct derivative of how we manage social interactions; all animal societies have coded social interactions, often of a very subtle nature.

Economics in a very fundamental way, is an extension of political or social control; always with particular benefits accruing to a select segment of society. The very words used to describe these relationships imply hierarchies of authority: employee, employer; worker, boss; CEO, manager; these (and many more) imply levels of authority and emoluments. What is totally missing is any sense of appreciation of human dignity in the sense that each of us feel pain when hurt; distress when we cannot feed our children; joy when stimulated by beauty, and a need to be appreciated.

Payment of a wage or salary for "work" done is convenient, but it does not say thanks and I hope you enjoyed the sunset. It implies don't spend this foolishly and be sure to be here on time tomorrow.

What I am leading up to is this; as a complex animal society we do not work within that society for the common good—"Some animals are more equal than others." No entrepreneur or CEO could have any economic benefits without the work of other human beings. These human beings are not "persons, goods, services and capital."

The linking of persons with "goods, services and capital" is implicitly denying them their basic human right of the dignity of being a free member of the human race. They have become an economic chattel. Those who have economic control will, I am sure, deny the above statement with great indignation. But what cannot be denied is that in fact we are all, the CEO included, subject to the economic structure we have created. None of us are free.

We need a constitution that recognizes this, and then tries to point a way to a goal that will eventually enable us to manage our society so that the whole of human activity is predicated on the idea that our collective activity has to provide, as a right, food, shelter, education (sharing of knowledge,) health care, and leisure to enjoy our own personal goals. (I am aware of the justiciable problems of entrenchment in the body of the Constitution, but this concept must eventually become part of the fabric of our society.)

After these had been provided for all, then we could share the rest according to our special skills and interests. But if the whole is not sufficient to supply our basic rights, then those rights must be provided for, before any extra emoluments are made available to any other segment of society.

What I have described is an extension of the Aboriginal potlatch. We have much to learn from our First Nations. Their respect for the wisdom of their elders, their administration of justice according to circumstance and service to the community, their use of consensus rather than political domination; their reverence for the land and its life—all of these which were developed over many thousands of years, stand in stark contrast to our wanton and crass methods of letting our greed and insensitivity govern our society.

Yet in fairness, I must say that we, in Canada, have done a better job than most nations. We can and must do better. The time is now.

I fully realize that much of what I have said above will be regarded as highly idealistic and perhaps naive. I firmly believe that most of us are so close to the urgency of our daily lives that it is difficult, and perhaps impossible, for us to look at Canada and the planet from Roberta Bondar's vantage point and to see it for what it really is—a precious and beautiful jewel set in the vast blackness of infinite space. It is ours to cherish; it is ours to understand; it is ours for the moment, in the future it will be the responsibility of others; it is unique, beautiful and absolutely essential to our lives. The ego and self-interest that drives so much of human behaviour is perhaps our planet's most serious threat. Let our Constitution set in place some control of that threat and point us onto a path that will nurture and cherish the sparkle of this jewel we call home.

CHAPTER TWENTY-ONE
WORLD-WIDE PROBLEMS, DEFICITS AND THE UNIVERSALITY OF EXCESSIVE INCOME—1992

These comments are not made by an economist! I have no formal training in economics; I have lived in and observed the world scene for over seventy years and these observations stem from that experience.

It would appear that deficit financing began when money was first used as a medium of exchange and human greed first started extracting usurious interest rates in return for the loan of money. The cost of borrowing money has been a factor in all economies for centuries. It is now such a well established practice that it is accepted as a fundamental factor in all fiscal policies; repayment for the use of borrowed money has resulted in the accepted practice of lending money (investing money) where it will generate the biggest return. International investors shift their money from country to country on, what is now, an almost minute-by-minute basis, thanks to modern communications technologies.

Whole industries are devoted to assessing the credit-rating of countries as a service to those who have huge sums of money to invest.

Canada's rating was recently reduced from AAA to AA+; generally taken as a sign of a weakened economy.

It appears to me that the world economy is bound so closely to mortgaging the present for increased productivity in the future, that no country can escape deficit financing if it wants to expand its economy and improve the living standard of its population. Since we live on a finite planet with finite natural resources we can only continue on this path by increasing our intellectual output, not by continued economic (read "industrial") growth.

The rate of technological advance seems to be outstripping our educational preparation for the new world we are creating. The rate of "bored" dropouts from our schools seems to be increasing; the pool of highly skilled talent is probably shrinking rather than growing. The listing of skilled positions in the week-end editions of newspapers indicates that this is indeed the case. Where then is the "intellectual output" going to come from in our finite physical world? Where then is the necessary increase in our monetary wealth going to come from to finance the ever growing deficits?

The inevitable consequence of continuing along this path will be world-wide economic collapse with tragic results. Unless the age-old practice of free enterprise coupled with variable (and usually high) interest rates is modified to comply with the new world reality, I can see no other end but disaster.

Let us now ask a few fundamental questions: To whom is all this money owed? How many persons (not corporate bodies) are in this group? Are they personally at risk if the system fails? Are their children at risk?

My guess is that the answers will depend upon to whom we put the question. It is, in my opinion, highly probable that the numbers are quite small; that they are unlikely to be personally at risk and that they do not have strong personal allegiances to any particular country.

Let us now ask a different set of questions: How many people are living in dire poverty? If they are not living in poverty, how many are unemployed and are dependent upon personal or state charity? Are they at risk if the system fails? Are their children at risk? There is no doubt about the answers. We see the answers all about us every day. Millions of people are at risk and nearly all children are at risk.

If the Darwinian theory of survival of the fittest applies in the world of economics, then the situation is quite normal and the human cost is not of any long-term significance. If, on the other hand, human compassion is a factor then the situation is an unmitigated tragedy. (*The Spirit Level—Why Equality Is Better for Everyone*, by Richard Wilkinson & Kate Pickett, addresses this problem with remarkable clarity.)

Having tried to outline the bare bones of my case, I now ask "What can we do about the problem?" Let me say right away that I do not see any hope of those within the system changing it. As I write, a documentary on CBC NewsWorld is examining the "Dachaland" of the Russian elite; one occupant said that he would burn his dacha rather than have it taken away from him. Highly privileged people do not give up their privileges easily.

I therefore am going to propose the changes I see as essential, even if they may not have any significant chance of becoming reality.

Deficits occur because governments borrow money on the premise that economic growth stimulated by the use of the borrowed money, will more than pay back the principal and interest. Other factors usually seem to interfere and the debt accumulates. New governments promise to solve the problem, but they too, borrow more money, and the debt again rises. In the last few decades we, in Canada, have accumulated a deficit of nearly $500 billion and the deficit is still increasing at about $30 billion per fiscal year. This is a no-win situation.

If it is a no-win situation, how do we ever get out from under this crushing burden? The usual answer is "Increase our productivity and cut government spending." But what actually happens? Political expediency gets in the way and governments continue in the old ways because they can neither see, nor devise, any effective way to solve their problems and still stay in power. (They can't burn their dachas!)

It seems to me that the only solution is for the world community to re-write the rules and start over again. When Shylock demanded his pound of flesh, he did not know about computers and modern, instantaneous, world-wide communications. The world we now live in bears almost no resemblance to that of the money lenders whom Jesus supposedly threw out of the temple. Yet his solution may, in fact, be the one we must now use.

Let us suppose that we repudiate ALL world debt. Those who are the "lenders' would have their livelihoods disappear overnight; but they are probably fewer in number than the children dying daily from starvation. Governments would have a clean slate from which to finance their country's economies. The crushing tax burden on the "people" that is required to finance the deficits would disappear. What would not disappear would be the moral imperative to never again operate under a system that created such enormous disparities in the human condition on this planet.

What should those conditions be? There is no doubt that the accumulation of both money and power would begin again immediately. What rules could we possibly invent to prevent, or at least control, this return to disaster?

Money would still be necessary for the world's commerce to flow; the use of money would still demand some recompense for that use; speculators would still try to gain any advantage they could from manipulation of the monetary market place.

Interest rates have traditionally been determined by both demand and government policy. These factors have been so variable that control is soon lost and the inevitable inequities arise. It seems to me that the only sure way to exercise any meaningful control is through a fixed, world-wide interest rate. If this were done by international agreement, then the conditions within any country would be imposed by that country's own decisions coupled with similar conditions and decisions elsewhere. In other words, any country would have to rely on its own resources and human potential for its progress and not on monetary decisions made by a comparatively few speculators seeking their own gain.

What should such internationally fixed and permanent interest rates be? The interest rate must be low; it must not allow personal (or corporate) control of large blocks of money to be used for purposes other than as a service supplied at nominal cost. It should be related to reasonable and acceptable real growth rates of both knowledge and material goods and services world-wide. Excessive rates of growth tend to produce enormous disparities, not only in societies, but also in inevitable pollution and environmental damage. We have begun to learn that lesson in recent years.

It seems that actual growth rates in excess of a few per cent become destabilizing. Therefore, interest rates should be coupled to those same growth rates. If this is a tenable hypothesis then interest rates should be no more than one to two per cent per year and permanently fixed, subject to periodic international adjustment, say every ten years, until real stability has been established.

[2012: I did not anticipate the extremely low interest rates of recent years!]

With permanent, low interest rates and the deficit wiped out world-wide, societies could realistically develop economies that provided for the well being of all peoples, rather than the tragic inequities that now plague most of the world's population.

Idealistic? Perhaps, but there seems to be no doubt, that if we continue on our present course the inevitable consequences will be truly disastrous and completely beyond our control. Almost all discussions on economic problems assume that the fundamental structure is a "given" and must not be tampered with, because of the sacred covenant that debt is sacrosanct and the lender must be paid in full. I don't agree. We can, indeed we must, change our monetary system, otherwise our children and those who succeed them will hold us accountable for a tragedy whose dimensions we cannot yet imagine.

Universality of Excessive Income

The recent Tory budget continued the attack of the concept of universality of basic social support systems in Canada. William Gold in The Calgary Herald (Sunday, March 1, 1992) has attacked the "loonie-left do-gooders" for criticising this aspect of the budget. It does not seem to occur to people that there might be another type of universality we might consider in our national agenda - that is universality as applied to "excessive" income, such as massive corporate profits, which are distributed to shareholders as dividends based upon shares held as equity in that corporation.

It seems to me that the fatal flaw in modern day capitalism is just those dividends. The inevitable result will be the bankruptcy of the entire fiscal world and a massive degradation of the world environment. Unless we restructure our economic world so that "profits" are properly fed back into a support system and are used to facilitate the necessary adjustments to ecologically responsible human behaviour, we will die in a disastrous economic soup of pollution.

There is nothing wrong with rewarding enterprise and ingenuity with a good standard of living; what is wrong is to use excessive profits to support a profligate standard of living while doing massive harm to both society and the environment.

The concept of universality as applied in this sphere would ensure a high standard of living for those whose intelligence and insights enables society to survive and prosper, and would mean that a basic support is provided for all, and where available, the rest is shared according to contribution, up to a limit defined by the needs of the rest of humanity. Such a concept is anathema to those of the "capitalist" persuasion.

But what are we really doing in this world. We have created and enshrined a system that accrues untold wealth and power to those whose drive and insensitivity have enabled them to write their own rules. Because they are successful they then are in a position to persuade the rest of us that we too can emulate them and also become wealthy. What they don't reveal is that those very rules are designed to prevent this from happening.

Back to the concept of universality. If everyone had to assume some responsibility for the viability of our society, then it would be entirely reasonable that universality would apply to every citizen without regard for status or wealth. The overall wealth would also be shared. The cry of discrimination, communism, socialism, destruction of "free enterprise" and other alarums would be deafening, but at least there would be a reasonable chance that there might still be a viable civilization on this planet in the year 3000 - otherwise I see little hope. The inhuman exercise of power that has produced such enormous inequities in the world would be at least abated. The assault on the environment would occur at a much more slow pace and we would have a chance to observe its effects and change course if necessary.

CHAPTER TWENTY-TWO
TV VIOLENCE AND THE SAFETY
OF CHILDREN—1993

It is increasingly apparent, [2012] from watching current TV programs, that the use of violence as entertainment has not decreased; the subject seems to have been relegated to the back burner. The current development in the neurosciences will soon enable the study of the development of young brains. I would not be surprised to find that the effect of TV violence, including electronic games, on young brains is a desensitizing of their empathy for others.

Recently, Prime Minister Mulroney took advantage of a "photo-op" to sign a petition deploring the effect of TV violence as a factor in the increasingly hazardous life of children. Parents have for many years been concerned with violence as portrayed on TV. The question is "What to do about TV violence?" Such petitions seem to have little real effect and, as Trudeau once said, "The State has no place in the Bedrooms of the Nation"—neither, it appears, does the State have any place in the Living Rooms of the Nation, where it can exercise insidious "thought control" of the population and its enterprises.

Yet TV violence is of real concern. Most people do not enjoy experiencing deliberate violence to themselves or their loved ones. The sociology of violence is another problem, but there is little doubt that violence in our society has become endemic.

The producers of TV entertainment shows, where violence plays a dominant role, produce such shows because they make money, not to enhance the culture of the country or to improve the quality of life for children. They therefore, as a business, should pay a tax for this privilege.

I propose a VIOLENCE TAX on all programs in the "entertainment" category shown on TV at any time. This tax would be based on three categories only; "x" dollars per physical assault incident; "5x" dollars per assault with willful property damage; and "50x" dollars where death results from willful violence. Collecting such a tax must not increase our over-bloated bureaucracy. I therefore propose that, before the program can be shown, the producer must file a Violence Index, based on these three categories and submit the appropriately calculated tax to the proper authority. The tax would be applicable each time the program is shown. The government's role would be to randomly monitor TV programs and administer an applicable, but severe fine, for non-compliance.

In the case of cable TV it should be the responsibility of the cable company to ensure that their clients comply with the law before signing contracts to show violent TV programs.

This process does not prohibit the production of violent TV shows. It simply taxes those who choose to make a living from cynical exploitation of violence between human beings.

I recognize the problem of establishing categories; for example, old movies made before the days of television should probably be exempt.

The problem of establishing the categorization of any particular act, as portrayed on TV, as an act of violence should not be subject to a Code of Ethics as devised by either a Regulating Authority or the

Entertainment Industry; rather the validity of the categorization should be *"Does this particular act violate the laws of the country in which it is portrayed?"*

In a Southam News article by Ian Austen appearing in the October 29, 1993 issue of the *Calgary Herald*, page A1, the following statistics were presented in a sidebar:

> 1,846 definable acts of violence were counted by *TV Guide* during an 18-hour period of TV programming on 10 different channels—they were categorized as follows:
>
> Other—17%
>
> Menacing threat 12%
>
> Pushing/dragging 15%
>
> Isolated punches—15%
>
> Gunplay—20%
>
> Serious assaults (no guns)—21%

Using the multipliers suggested above, this amounts to 21,487 "x" dollar incidents in any 18-hour period. If one assumes a modest violence tax of $100 per incident, this then translates into a Violence Tax revenue of $2,148,700 per day. Surely this would give pause to those who choose to make their living by degrading our society by exploiting violence. If this tax value did not reduce TV violence, then it could be increased to the point where it would be effective. (As well as affective!) I expect that the threat of such a tax might be quite a strong incentive for reform!

Note that this proposal does not say that violence may not be used in any TV program. It simply says that if one wants to depict violent acts, which are in contravention of the laws of our country, then they must pay a tax for that privilege.

The argument that no actual violence was perpetrated by depicting it in simulated form on TV and therefore it is not in violation of the laws of the land, while technically true, is not ethically valid— this fine point most certainly is not understood by many children. The increasing violence in our schools, and indeed throughout our society, is indicative of a serious breakdown and eventual collapse of a humane and civilized society in Canada—the current concern about bullying for example.

The proposed "blackout" track on VCR cassettes, while effective if used, applies to a rather limited segment of the entertainment industry (Disney Studios may have devised such a track to prevent unauthorized dubbing of their VHS videos.) It certainly would not prevent "latchkey" children from watching violence on TV, especially in the coming communications revolution.

Such a tax would probably have an indirect influence on the use of violence in movies produced for movie theatres, where people have to go in person to see the entertainment.

Violence in our society, to a much larger extent, is due to unemployment and the social degradation of being on welfare. Also, TV violence has more influence in this segment of our society because they tend to use TV as an escape from the problems of their everyday lives, and for the same reason it has a profoundly negative influence on their children.

The central point of my argument is that we in Canada, as citizens, are in fact agreed through our laws, that violence is not to be tolerated in our society without penalty, and that if the quality of life is degraded, by any means, in contravention of those laws, there should be a penalty for so doing.

President Clinton will probably not be hosting any sleepovers in Lincoln's bedroom for Hollywood's elite in the near future. Indeed, in lashing out at the entertainment industry, Clinton has turned his back on his rich and famous West Coast buddies, the very people who have poured millions into the Democratic party and his campaign coffers. In a June 1 Rose Garden speech, the president announced plans to commission a study to investigate whether violent films, television shows, video games, and music are marketed toward children, specifically young boys. In effect, Clinton is trying to determine if the entertainment industry is setting its sights on the impressionable and vulnerable segment of society the same way the tobacco industry has. The Justice Department and the Federal Trade Commission will conduct the yearlong study.

> "They, and the rest of us cannot kid ourselves," Clinton said. "Our children are being fed a dependable daily dose of violence—and it sells."

The Government's response to this idea was completely negative; I was not surprised!

CHAPTER TWENTY-THREE
POLITICIANS & THE SCIENCE OF CLIMATE CHANGE—2011

"Climate Change" as it is discussed in the media and in government circles as well as by the "public" bears little relevance to the physics and chemistry of the actual phenomena. The term Climate Change was introduced by agents of the major petroleum producers to reduce the impact of the correct terminology, namely, Global Warming (GW).

For most politicians, Climate Change has become a concept that is subject to man-made laws of governance and therefore can be dealt with in due course, or not at all.

The warming effect of CO_2 in the Earth's atmosphere has been known for more than 100 years. That effect had become of concern to knowledgeable scientists and organizations in many countries; it resulted in the Kyoto Protocol, which defined the problem and the necessary actions required to prevent the level of CO_2 from rising to the tipping point of 850ppm, when global warming would spiral out of our control.

This portion of the Kyoto Protocol as presented to the United Nations, and adopted in 1997 at Kyoto, Japan was obtained from the Wikipedia web site:

"The Kyoto Protocol is a protocol to the United Nations Framework Convention on Climate Change (UNFCCC or FCCC), aimed at fighting global warming. The UNFCCC is an international environmental treaty with the goal of achieving the stabilization of greenhouse gas concentrations in the atmosphere at a level that would prevent dangerous anthropogenic interference with the climate system.

The Protocol was initially adopted on 11 December 1997 in Kyoto, Japan, and entered into force on 16 February 2005. As of September 2011, 191 states have signed and ratified the protocol. The only remaining signatory not to have ratified the protocol is the United States. Other states yet to ratify Kyoto include Afghanistan, Andorra and South Sudan, after Somalia ratified the protocol on 26 July 2010.

Under the Protocol, 37 countries (Annex I countries) commit themselves to a reduction of four greenhouse gases (GHG) (carbon dioxide, methane, nitrous oxide, sulphur hexafluoride) and two groups of gases (hydrofluorocarbons and perfluorocarbons) produced by them, and all member countries give general commitments. Annex I countries agreed to reduce their collective greenhouse gas emissions by 5.2% from the 1990 level. Emission limits do not include emissions by international aviation and shipping, but are in addition to the Industrial gases, chlorofluorocarbons, or CFCs, which are dealt with under the 1987 Montreal Protocol on Substances that Deplete the Ozone Layer.

The benchmark 1990 emission levels accepted by the Conference of the Parties of UNFCCC (decision 2/CP.3) were the values of "global warming potential" calculated for the IPCC Second Assessment Report. These figures are used for converting the various greenhouse gas emissions into comparable CO2 equivalents (CO2-eq) when computing overall sources and sinks.

The Protocol allows for several "flexible mechanisms", such as emissions trading, the clean development mechanism (CDM) and joint implementation to allow Annex I countries to meet their GHG emission limitations by purchasing GHG emission reductions credits from elsewhere, through financial exchanges, projects that reduce emissions in non-Annex I countries, from other Annex I countries, or from Annex I countries with excess allowances.

Each Annex I country is required to submit an annual report of inventories of all anthropogenic greenhouse gas emissions from sources and removals from sinks under UNFCCC and the Kyoto Protocol. These countries nominate a person (called a "designated national authority") to create and manage its greenhouse gas inventory. Virtually all of the non-Annex I countries have also established a designated national authority to manage its Kyoto obligations, specifically the "CDM process" that determines which GHG projects they wish to propose for accreditation by the CDM Executive Board."

Note that Canada ratified the Protocol but has ignored that signed commitment and established its own GHG (Green House Gas

Emissions) program which does not meet the Kyoto Protocol. Canada (the current Harper government) argues that all countries must adhere to an established collective agreement first; which is an excuse to do nothing that will harm Canada's economic concerns! That is, Harper's economic concerns.

Mike De Souza, POSTMEDIA NEWS, in *The Montreal Gazette* on November 29, 2011 wrote, in part:

> "Nearly three dozen countries, including Canada, Japan and Russia, took on targets in the first phase of the agreement, which is based on a principle in the convention that developed countries are responsible for causing climate change and must act first to address the problem.
>
> Harper and several members of his cabinet and caucus have previously questioned the scientific evidence linking human activity to global warming, describing the Kyoto Protocol as a "socialist scheme."
>
> But [Environment Minister Peter] Kent said Canada's anti-Kyoto stance was a key to a larger global agreement that requires the biggest annual sources of greenhouse gas emissions, such as the United States and China, to also take on targets that would stabilize emissions in the atmosphere and help the world avoid warming of more than 2° C above pre-industrial average temperatures—considered to be a dangerous threshold that could cause irreversible damage to the planet's ecosystems."

Shortly after winning his minority government, Harper abolished the PMO position of Science Advisor. Apparently, Harper thinks that Canada's problems are solely economic and that science is not a necessary ingredient in formulating plans for Canada's future.

Stephen Harper and his cabinet, as well as his caucus, do not realize that no matter what they think or do, Nature, that is the entire natural universe, neither knows nor cares, but will react according its own immutable laws. That is why a creditable science advisor is essential in modern times.

The current scientific understanding of the physics of the presence of certain gasses in the Earth's atmosphere is that the average global temperature is rising, and will continue to rise at an increasing rate as long as either, or both, methane and CO_2 (the most prevalent GGs) are added. Further, we are approaching the situation where the atmospheric warming has become exponential and will soon become unstable and increase beyond our control.

Global warming occurs because light radiation from the sun passes through the atmosphere with little effect, however, the infrared radiation from a warmed earth reflected back through the atmosphere does warm the Greenhouse Gases in the atmosphere; once warmed, CO_2 slowly loses its heat over an extended period approaching 100 years. CO_2 is also slowly absorbed by the oceans, increasing their acidity. The melting of the polar ice caps reduces the reflecting ice surface as well as exposing more ocean water surface to solar radiation, thus accelerating the increase of the overall global temperature.

This very simplified explanation of global warming will suffice to indicate that it is a matter of physics and chemistry, not politics. (One wonders if politicians don't make their own contribution to global warming!)

In 1967, I delivered a lecture in which I said: "…but never suffer the delusion that Nature will alter her ways because we do not understand her and that by sheer reason or logic we can construct a framework of belief that will stand against the simplicity, elegance and beauty of an absolutely obdurate Nature." (I was unaware at that time of the role of CO_2 in the atmosphere.)

"Science, that branch of human behaviour which fundamentally seeks only to discover the truth, and technology, the art, usually industrial, of applying the fruits of scientific research in a manner which makes them of practical use in the broader field of human society, have been so misapplied during the past (six) years that thinking men have grave doubts for the future of our present civilization."

This brief paragraph from an earlier talk of mine in 1946 is still valid, and illustrates a deficiency in the our general education of all citizens—politicians in particular—because of the essential need for everyone to understand the inevitable consequences of our political decisions and actions.

When Prime Minister Harper remarked that the Kyoto Protocol was a "socialist scheme" he was demonstrating his deplorable lack of knowledge of the physical world.

Following the *Montreal Gazette* article was a Comments column; there were 60 entries. Thirty-nine indicated that the writer had no concept of the seriousness of GW; 11 were concerned; 10 were ridiculous or just confused. This unscientific analysis, plus many opinion polls, and the Canadian government's reaction to the Global Warming meeting in Durban, South Africa, tells me that we, here in Canada, have lost any credibility we once had on the world stage.

I did hear Elizabeth May (the sole elected member of the Green Party) speaking in the House and more recently on the Parliamentary Channel CPAC and *CBC Newsworld*; she clearly understood the causes of GW and the urgent need to curb the burning of fossil fuels.

I have an additional concern. We must support a massive research effort to devise and develop alternative energy sources which do not depend on fossil fuels. Such an effort will eventually be rewarded with hope for us and our children's future, as well as the creation of wealth and jobs. During the National Research Council's golden years (1950-1970) there was an active Section examining the problem of access to solar radiation as a source of energy to power the world's industries; it was disbanded—I don't know what authority made that decision. Currently, a major portion of the profits from the exploitation of Canada's tar sands should be used for exploring a method of utilizing solar radiation—the total amount available to us on the Earth's surface far exceeds our future needs.

READING LIST FOR FURTHER READING ON CLIMATE CHANGE

Age of American Unreason, The, by Susan Jacoby—Pantheon Books, New York, 2008

Assault on Reason, The, by Al Gore—Penguin Books, New York, 2008

Big Coal, The—Dirty Secret Behind America's Energy Future, by Jeff Goodell—Hougton Mifflin Co., Boston, 2006

Carbon Shift—How Peak Oil and the Climate Crisis Will Change Canada—Editor, Thomas-Homer-Dixon—Random House of Canada Ltd., Toronto, 2010

Climate Culture Change—Inuit and Western Dialogues with a Warming North, by Timothy B. Leduc—University of Ottawa Press, 2010

Climate War, The—True Believers, Power Brokers, and the Fight to Save the Earth, by Eric Pooley—HarperCollins, New York, 2010

Global Carbon Cycle, The, by David Archer—Princeton University Press, Princeton, 2010

Heat—How to Stop the Planet from Burning, by George Monbiot—Doubleday Canada, Toronto, 2006

Heatstroke—Nature in an Age of Global Warming, by Anthony D. Barnosky—Island Press, Washington, DC, 2009

Hot—Living Through the Next Fifty Years on Earth, by Mark Hertsgard—Houghton Mifflin Harcourt Publishing Co., New York, 2009

Hot Air, Meeting Canada's Climate Change Challenge, by Jeffrey Simpson, Mark Jaccard and Nic Rivers—McClelland & Stewart Ltd., Toronto, 2007

Storms of My Grandchildren—The Truth About the Coming Climate Catastrophe and Our Last Chance to Save Humanity, by James Hansen, Bloomsbury, USA, 2009

Unscientific America—How Scientific Illiteracy Threatens Our Future, by Chris Mooney and Sheril Kirshenbaum—Basic Books, New York, 2009

Weather Makers, The, How We Are Changing the Climate and What It Means for Life on Earth, by Tim Flannery—HarperCollins Publishers Ltd., Toronto, 2006

POSTSCRIPT

The content of this book spans nearly seventy years. For most of that time I had hopes that I would see changes that would lead to a promise of a brighter future for my grandchildren. That hope is fading. Recent Canadian Federal Governments have betrayed the vision of the Founding Fathers. Honesty and integrity do not seem to be in their arsenal of qualities. In particular, our present government is edging toward fascism in its total disregard for the wishes of the majority of Canadian people. Moreover, they have assumed the righteousness of religious fanaticism in their policy making.

We desperately need to fashion a new mode of governance. Democracy and capitalism are incompatible—the professed beliefs of a greatly divided religious community are not helpful; which reminds me of George Orwell's book Animal Farm and this famous quote; All animals are equal, but some animals are more equal than others. This is exactly why capitalism and democracy are incompatible.

Perhaps you and your friends can find that new pathway into the future.

ACKNOWLEDGMENT

I could not have succeeded in this endeavour without those 57 years of fond encouragement by Mavis and the unstinted support of our children, Margaret, John, David and Daniel.

ABOUT THE AUTHOR

McNarry was born in a farm house in South-West Manitoba in 1916. Following service with the RCAF while attached to the Royal Air Force as a Radar Officer, he obtained his Master's Degree in Mathematics and Physics at The University of Western Ontario and was subsequently employed as a research scientist with The National Research Council in Ottawa. He was nominated for the Marconi International Fellowship Award in 1981. He was an occasional consultant to the Ontario Department of Education in development of science courses following his retirement in 1979.

Lightning Source UK Ltd.
Milton Keynes UK
UKOW04f2010230913

217779UK00002B/611/P